# Development and Applications of
# Free Electron Lasers

China Center of Advanced Science and Technology (World Laboratory)
Symposium/Workshop Proceedings

Editors in Chief

WORLD LABORATORY
**A. Zichichi**

CCAST
**T.D. Lee**
**K.C. Chou**

# Development and Applications of Free Electron Lasers

*Edited by*

**Jia-er Chen**
*Peking University*
*Beijing*
*People's Republic of China*

**Hai-cang Ren**
*The Rockefeller University*
*New York, USA*

**Jia-lin Xie**
*Institute of High Energy Physics*
*Chinese Academy of Sciences*
*Beijing*
*People's Republic of China*

**Ming-han Ye**
*CCAST (World Laboratory)*
*and*
*Institute of High Energy Physics*
*Chinese Academy of Sciences*
*Beijing*
*People's Republic of China*

Proceedings of the CCAST (World Laboratory)
Symposium/Workshop
held at
China Center of Advanced Science and Technology (World Laboratory)
*Beijing, People's Republic of China*

May 29–June 3, 1995

**Gordon and Breach Science Publishers**

Australia • Canada • China • France • Germany • India • Japan
Luxembourg • Malaysia • The Netherlands • Russia • Singapore • Switzerland
Thailand • United Kingdom

Amsteldijk 166
1st Floor
1079 LH Amsterdam
The Netherlands

British Library Cataloguing in Publication Data

Development and applications of free electron lasers. –
  (China Center of Advanced Science and Technology (World
  Laboratory) Symposium/Workshop proceedings, v. 12)
  1. Free electron lasers – Congresses   2. Free electron lasers
  – Industrial applications – Congresses
  I. Chen, J.   II. CCAST (World Laboratory) Symposium/ Workshop
  (1995 : Beijing, China)
  621.3'66

ISBN  90-5699-502-2

# Contents

# 丛 书 前 言

中国高等科学技术中心(CCAST)于 1986 年 10 月在北京成立。它是一个民间研究组织,它参加由意大利政府捐助建立的民间组织世界实验室并获得世界实验室的资助。它和中国的研究机构及大学建立密切的合作关系。中心的目的是在中国创造良好的研究环境,建立中国和世界各国研究机构和科学家之间的联系,以鼓励中国科学家做出有世界水平的研究成果,为全世界科学技术自由交流和发展做出贡献。

中心的主要活动之一是组织国际学术交流:每年将组织三至四次国际学术讨论会,精选与中国现有条件相适合的和可发展的前沿科技领域,邀请在这方面有突出贡献的科学家,介绍该学科的基础、现状、特别强调存在的问题和困难及将来可盼望的前景,使之有助于中国及世界的物理学家共同努力,推动这一学科的发展。

每次讨论会将有一本文集,这部丛书就是这些文集的汇合。希望中国的物理学界尤其是年轻的一代能从这部丛书中得到教益。

李政道

# Preface to the Series

The China Center of Advanced Science and Technology (CCAST) was established in Beijing on October 17, 1986 to introduce important frontier areas of science to China and to promote the free exchange of scientific information between China and other nations. It is sponsored by the World Laboratory, with support from the Italian and Chinese governments.

Every year CCAST (World Laboratory) organizes three or four international symposia/workshops on subjects that are especially selected for their potential as seeds for future development in China. Each symposium brings together about 10 experts from abroad and 60–70 scientists from within China. They work very closely to discuss, in depth, the state of the art of the subject and to explore its future possibilities, with special emphasis on present problems within the area. It is this joint labor of fostering the growth of modern science in an ancient center of civilization that gives these symposia an especially uplifting feeling. This series of proceedings may serve as witness to these efforts on behalf of the younger generation of Chinese physicists.

T.D. Lee

# Invited Lecturers

D. Oepts — FOM-Instituut voor Plasmafysica "Rjinjuizen", Edisonbaan 14, 3439 MN Nieuwegein, The Netherlands

Claudio Pellegrini — Physics Department, University of California, Los Angeles, CA 90095, USA

Richard L. Sheffield — Los Alamos National Laboratory, MS H851, Los Alamos, New Mexico 87544, USA

Norman H. Tolk — Center for Molecular and Atomic Studies at Surfaces, Department of Physics and Astronomy, Vanderbilt University, Nashville, Tennessee 37235, USA

Ming Chang Wang — Shanghai Institute of Optics and Fine Mechanics, Academia Sinica, P.O. Box 800211, Shanghai, People's Republic of China

Jia-lin Xie — Institute of High Energy Physics, Chinese Academy of Sciences, P.O. Box 918, Beijing 100039, People's Republic of China

# 参加者名单
## List of Participants

| | | |
|---|---|---|
| Chen, Jiaer | 陈佳洱 | Peking University |
| Chen, Ming | 陈　明 | Shanghai Institute of Nuclear Study |
| Chen, Nengkuan | 陈能宽 | Institute of Applied Physics & Computational mathematics |
| Chen, Yu | 陈　煜 | Peking University |
| Dai, Zhimin | 戴志敏 | Shanghai Institute of Nuclear Study |
| Ding, Meisong | 丁梅松 | China Institute of Atomic Energy |
| Ding, Wu | 丁　武 | Institute of Applied Physics & Computational Mathematics |
| Dong, Zhiwei | 董志伟 | Institute of Applied Physics & Computational Mathematics |
| Du, Xiangwan | 杜祥琬 | Institute of Applied Physics & Computational Mathematics |
| Feng, Sunqi | 冯孙齐 | Peking University |
| Fu, Shinain | 付世年 | China Institute of Atomic Energy |
| Geng, Rongli | 耿荣礼 | Peking University |
| Guan, Genfa | 关根法 | Peking University |
| Guo, Kangzhu | 郭康杜 | Institute of High Energy Physics |
| Hei, Dongwei | 黑东炜 | Xibei Institute of Nuclear Technology |
| Hu, Kesong | 胡克松 | China Institute of Engineering Physics |
| Hu, Qiheng | 胡启恒 | Chinese Academy of Sciences |
| Hu, Suxing | 胡素生 | Shanghai Institute of Optics and Fine Mechanics |
| Huang, Xiaohai | 黄哓海 | Institute of High Energy Physics |
| Huang, Yu | 黄　羽 | Shanghai Institute of Optics and Fine Mechanics |
| Jia, Qika | 贾启卡 | University of Science & Technology of China |
| Jiang, Yunqing | 姜云卿 | Institute of Applied Physics & Computational Mathematics |

| | | |
|---|---|---|
| Lee, T.D. | 李政道 | Columbia University |
| Li, Dazhi | 李大治 | Shanghai Institute of Optics and Fine Mechanics |
| Li, Fengtian | 李逢天 | Institute of High Energy Physics |
| Li, Ge | 李 格 | University of Science & Technology of China |
| Li, Qinglu | 李清胪 | University of National Defense |
| Li, Yonggui | 李永贵 | Institute of High Energy Physics |
| Li, Yunjun | 李运钧 | Zhengzhou University |
| Li, Zhenghong | 李政红 | China Institute of Engineering Physics |
| Liang, Rongji | 梁荣基 | Chinese Academy of Sciences |
| Liang, Zheng | 梁 正 | University of Electronic Science & Technology |
| Liu Qingxiang | 刘庆想 | Chengdu P.O. Box 517 |
| Liu, Changan | 刘长安 | Xibei Institute of Nuclear Technology |
| Liu, Fengying | 柳凤英 | China Institute of Atomic Energy |
| Liu, Jie | 刘 洁 | Institute of Applied Physics & Computational Mathematics |
| Liu, Weiren | 刘慰仁 | China Institute of Atomic Energy |
| Lu, Huihua | 陆辉华 | Institute of High Energy Physics |
| Mitrochenko, Victor | 维克多 | Harkov Institute of Technical Physics |
| Oepts, Dick | 欧 泊 | FOM-Instituut voor Plasmafysica "Rijnhuizen" |
| Pan, Yuli | 潘雨力 | China Institute of Atomic Energy |
| Pang, Yang | 庞 阳 | Columbia University |
| Pelligrini, Claudio | 潘立格莱尼 | University of California, Los Angeles |
| Ren, Haicang | 任海沧 | Rockefeller University |
| Sheffield, Richard | 谢菲尔德 | Los Alamos National Laboratory |
| Shi, Yijin | 施义晋 | China Institute of Atomic Energy |
| Shu, Xiaojian | 束小建 | Institute of Applied Physics & Computational Mathematics |
| Su, Jing | 苏 憬 | Institute of High Energy Physics |
| Sun, Weilin | 孙伟林 | China Science and Technology Daily |
| Tang, Longzhou | 汤龙舟 | Chengdu P.O. Box 517 |
| Tolk, Norman H. | 托 克 | Vanderbilt University |
| Wang, Bosi | 汪伯嗣 | Institute of High Energy Physics |
| Wang, Chuilin | 王垂林 | CCAST |
| Wang, Daheng | 王大珩 | Chinese Academy of Sciences |
| Wang, Gang | 王 钢 | Institute of High Energy Physics |
| Wang, Ganchang | 王淦昌 | China Institute of Atomic Energy |

| Wang, Jianwei | 王建伟 | Institute of High Energy Physics |
| Wang, Lifang | 王莉芳 | Peking University |
| Wang, Mingchang | 王明常 | Shanghai Institute of Optics and Fine Mechanics |
| Wang, Mingkai | 王铭凯 | Institute of High Energy Physics |
| Wang, Tong | 王彤 | Peking University |
| Wang, Yanshan | 王言山 | Institute of High Energy Physics |
| Wang, yougong | 王友恭 | China Scientific News |
| Wang, Youzhi | 王友智 | Institute of High Energy Physics |
| Wang, Zihua | 王子华 | Institute of High Energy Physics |
| Weng, Zili | 翁自立 | China Institute of Atomic Energy |
| Wu, Gang | 吴钢 | Institute of High Energy Physics |
| Wu, Yuanming | 吴元明 | Institute of High Energy Physics |
| Xia, Jialin | 谢家麟 | Institute of High Energy Physics |
| Xu, Jin | 徐金强 | Institute of High Energy Physics |
| Xie, Xiaoyan | 谢小彦 | Peking University |
| Xu, Zhou | 许州 | China Institute of Engineering Physics |
| Yang, Guilin | 杨贵林 | China Institute of Engineering Physics |
| Yang, Zhenhua | 杨震华 | Institute of Applied Physics & Computational Mathematics |
| Ye, Minghan | 叶铭汉 | CCAST |
| Yin, Runhai | 应润海 | Institute of High Energy Physics |
| Yu, Junsheng | 俞俊生 | Xinan Institute of Physics |
| Yuan, Bin | 袁斌 | Zhengzhou University |
| Zhang, Baocheng | 张保澄 | Peking University |
| Zhang, Genshen | 张根深 | University of National Defense |
| Zhang, Lingyi | 张令翊 | Institute of High Energy Physics |
| Zhang, Shichen | 张士琛 | China Institute of Atomic Energy |
| Zhang, Yuzhen | 张玉珍 | Institute of High Energy Physics |
| Zhao, Kui | 赵夔 | Peking University |
| Zhao, Qiang | 赵镪 | Peking University |
| Zhao, Xiaofeng | 赵小彦 | Shanghai Institute of Nuclear Study |
| Zheng, Jingyun | 郑靖云 | Institute of High Energy Physics |
| Zhou, Wenzhen | 周文振 | China Institute of Atomic Energy |
| Zhuang, Jiejia | 庄杰佳 | Institute of High Energy Physics |

# HIGH BRIGHTNESS BEAMS AND APPLICATIONS

## RICHARD L. SHEFFIELD

Los Alamos National Laboratory, MS H851, Los Alamos, NM 87544 *

**Abstract** This paper describes the present research on attaining intense bright electron beams. Thermionic systems are briefly covered. Recent and past results from the photoinjector programs are given. The performance advantages and difficulties presently faced by researchers using photoinjectors are discussed. The progress that has been made in photocathode materials, both in lifetime and quantum efficiency, is covered. Finally, a discussion of emittance measurements of photoinjector systems and how the measurement is complicated by the non-thermal nature of the electron beam is presented.

* Work supported by Los Alamos National Laboratory Directed Research and Development under the auspices of the United States Department of Energy.

## 1. INTRODUCTION

This series of talks covers the generation of high-brightness electron beams and the associated accelerator structures. Our work on high-brightness accelerators was motivated by the need to directly produce high quality electron beams. Damping rings can be used to produce high brightness beams, but cost and complexity inhibit wide deployment of damping ring technology. Also some of the electron beam applications require few-picosecond long pulses, shorter than can be produced with damping rings.

Several accelerator applications require high-charge, high-quality electron beams. Free-electron lasers (FELs) are widely tunable, high-power sources of light capable of generating wavelengths not accessible to conventional laser systems. The design of many FELs requires high peak currents and high brightness beams for high gain. Another application is the reduction in cost or elimination of damping rings. A reduction in cost is possible because a brighter electron is used for injection into the ring. With a brighter beam the ring acceptance can be reduced with a consequent reduction in cost. One other application is Compton back-scattering. For efficient Compton back-scattering the electron beam must be focused into a small volume again necessitating a high-brightness beam.

A brief outline of this paper follows. After a general introduction to basic electron physics, we will compare different approaches to the generation of electron beams. For very high-brightness high-charge electron beams, the photoinjector appears to be the appropriate electron source. Following the discussion of electron sources, a description of the overall design of a high-brightness accelerator is presented. The high-brightness accelerator uses a technique of emittance reduction called emittance compensation This is followed by a review of an accelerator system, named the Advanced Free-Electron Laser (AFEL), based on the preceding design work. The AFEL section will cover design, construction, operational characteristics, and experimental results. The subsequent section covers the measurement of emittance using the quadrupole scan technique. Although the quad scan is a commonly used technique for measuring emittance, in a photoinjector-based system this technique can lead to erroneous emittance measurements. Finally a few of the possible applications of an AFEL system are presented.

## 2. PHASE SPACE AND EMITTANCE

To understand the performance of an electron machine, we need to quantify the quality of the electron beam. This quantification is done using the concepts of phase space and emittance.

Phase space is the basic tool by which charged particle beam transport is characterized. Phase space is typically represented by a two-dimensional plot of x' versus x, where x is the transverse rms radius and x' is the particle's angle with respect to the optic axis,

$$x' = dx/dz = (\gamma mdx/dt)/(\gamma mdz/dt) = p_x/p_z, \qquad (2.1)$$

where m is the electron mass, and $\gamma$ is the relativistic gamma, $p_x$ is the transverse momentum and $p_z$ is the forward momentum.

Plots of the phase space of a zero transverse-temperature and zero energy-spread beam are schematically represented in the left figure of Figure 1(a). The optical-ray trace equivalent to those ray-trace plots is schematically shown in the right figure of Figure 1(a). As can be seen in the phase space plots in Figure 1(a), a line with a negative slope represents a converging beam, a vertical line represents the beam at a focus, and a horizontal line represents a beam with a focus at infinity.

Beams and the beamline optics are not perfect. As a consequence, beams cannot be focused to infinitely small spots (uncertainty principle aside). Figure 1(b-e) gives a few of many effects that can impact the focusability of an optical beam. An electron beam would have analogous effects in either its phase space or its equivalent optics. The spot size of a beam as measured with a diagnostic is the projection of the beam distribution on the x-axis. The beam at its focus is defined to have the smallest extent along the x-axis.

For thermalized beams, as shown in Figure 2, or a distribution that does not have a recoverable correlation in phase space, a good measure of the focusability of a beam is the rms emittance of the beam.[1] The rms normalized emittance, $\varepsilon_n$, is calculated from the rms emittance, $\varepsilon_x$. The emittances are calculated using the equations,

$$\varepsilon_n = \beta\gamma\varepsilon_x = \pi\beta\gamma(<x^2><x'^2> - <x\bullet x'>^2)^{1/2}, \qquad (2.2)$$

where $\beta$ is the particle velocity divided by the speed of light and x and x' are the particle's transverse coordinate and angle of divergence from the optic axis, respectively, and $<>$ means an average over the electron distribution f(x,y,z):

$$<x^2> = \frac{\int \int \int f(x,y,z) \, x^2 \, dx \, dy \, dz}{\int \int \int f(x,y,z) \, dx \, dy \, dz} . \qquad (2.3)$$

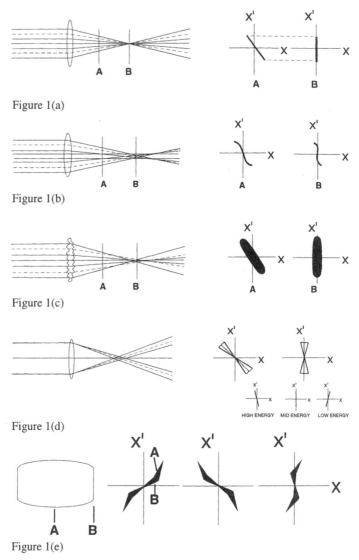

Figure 1(a)

Figure 1(b)

Figure 1(c)

Figure 1(d)

Figure 1(e)

Figure 1. In the left plot in (a), a collinear optical beam impinges on a perfect optic, passes through position A and is focused at position B. In the right plot in (a), the phase space plots corresponding to the two locations of A and B are given. In (b), the focusing optic has an incorrect curvature. In (c), the focusing optic's surface is rough. In (d), a beam composed of three energies (or wavelengths in the case of light) is focused. In (e), the first phase plot is a drifting beam expanding under the influence of space charge. The longitudinal middle of the beam at position A has more charge than at position B. The beam goes through a lens in the second phase plot in (e), and the third phase plot is the beam at a focus after the lens.

Beam dynamics are determined by transport of the phase space ellipse using matrix representations of the beamline optics. Either of two approaches are used to described the beam ellipse, the Twiss parameter or sigma matrix representation.[2] Figure 3 is a schematic showing the relationship between the Twiss and sigma matrix parameters.

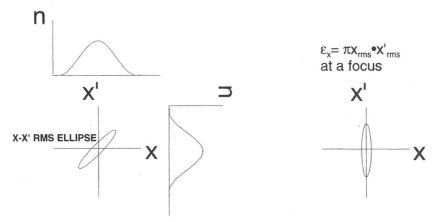

Figure 2. The area of an ellipse calculated from the rms x and rms x' values equal the rms emittance, $\varepsilon_x$, of a beam. This area is easily calculated at the beam's focus.

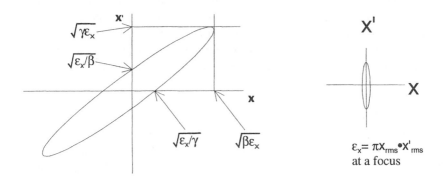

Eq. for using Twiss parameters: $\gamma x^2 + 2\alpha xx' + \beta x'^2 = \varepsilon_x, \ \beta\gamma - \alpha^2 = 1$

Eq. for using sigma parameters: $\sigma_{22}x^2 - 2\sigma_{21}xx' + \sigma_{11}x'^2 = \varepsilon_x^2, \ \sigma_{22}\sigma_{11} - \sigma_{12}^2 = \varepsilon_x^2$
where $\sigma_{22} = \varepsilon_x\gamma$, $\sigma_{11} = \varepsilon_x\beta$, and $\sigma_{12} = \sigma_{21} = -\varepsilon_x\alpha$.

Figure 3. Description of the Twiss parameters and sigma matrix parameters used for beam transport simulations. Note that $\beta$ and $\gamma$ are not the relativistic parameters.

Louiville's theorem states that the 6 dimensional phase space density is invariant.[3] However, because of the manner in which most experimental diagnostics work and the techniques that the computer simulations use to calculate emittance, only a projection of the phase space ellipse onto x-x' (or y-y' or r-r') plane is calculated. An emittance calculated from a projection of phase onto a plane is not a conserved quantity. This impacts the emittance experiments described later in this paper.

## 3. ELECTRON SOURCES

### 3.1 Thermionic Sources

To have a high-brightness electron beam, either the electron gun system must directly produce a bright beam or a damping ring must be used. Electron guns using a long pulse or a dc beam rely on a well-designed gun producing a beam that has a beam temperature near the thermal limit of the electron source. The beamline design after the gun depends on whether the application ultimately requires a dc beam or a short pulse. For a dc beam (or pulsed beams where the pulse end effects are negligible), very good quality beams can be produced if care is taken in the beam transport design. If the application requires a short pulse, then a bunching system must be designed that preserves the beam quality throughout the bunching and acceleration process. Preserving beam quality is difficult because of the effects of nonlinear rf fields in the bunching cavities and the space-charge forces present at sub-relativistic energies.

#### 3.1.1 DC guns

DC guns can produce beams with an emittance near the thermal limit determined by the cathode. The lower limit of the beam's normalized emittance from a thermionic electron source is governed by the emitter size and by the transverse component of the thermal motion of the electrons. The thermal limit of the normalized rms emittance of a beam from a thermionic emitter of radius $r_c$ at a uniform absolute temperature T is

$$\varepsilon_n = 0.5\pi r_c (kT/m_o c^2)^{1/2} \quad \text{[units: m · rad]} \quad (3.1)$$

because $<x \cdot x'> = 0$ at the cathode.[4] For a typical thermionic emitter at 1160 K, the average transverse energy of emitted electrons is 0.1 eV. For a uniform current density J, the total current is $I = \pi r_c^2 J$ and the lower limit on the rms emittance is

$$\varepsilon_n = 1.25 \times 10^{-6} (I/J)^{\frac{1}{2}} \; \pi \cdot mm \cdot mrad, \qquad (3.2)$$

with J in A/cm².

The current density from a dispenser cathode is typically no more than 20 A/cm². For example, the lowest achievable rms emittance for 1 cm² thermionic cathode is 0.28 $\pi$ mm-mrad.

The following information on very long pulse (>> 1 ns) and dc injectors is a summary of a paper[5] by W. Herrmannsfeldt. These types of guns are well suited for two applications: first, for electron cooling of ion beams and, second, for electrostatic free electron lasers (FEL). The design of a DC gun must include the effects of space charge. Figure 4 is a schematic of a DC gun designed for the University of California at Santa Barbara.[5] In the gun, the space-charge self-force in the beam is canceled out with a carefully designed focusing electrode at the Pierce angle,[6] thus maintaining a uniform current density. Also, the exit energy of the beam from the gun should be as high as possible to minimize further space charge defocusing downstream from the gun. If the beam maintains a uniform profile up to relativistic energies, then the beam emittance can be near the cathode-limited thermal temperature.

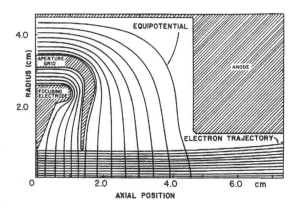

Figure 4. DC gun designed for UCSB.

Unfortunately, many applications require high charge per bunch in short pulses. The process by which the beam is produced and injected for acceptance into rf accelerator structures causes the emittance to grow by at least an order of magnitude. Also, high repetition rates (>1 MHz) are difficult. As stated above, bunches can be accelerated with a dc field and not suffer the emittance growth that is due to time-varying effects typically found in rf accelerators. The addition of harmonics to the rf accelerating fields has been proposed to eliminate this source of emittance growth. A design[7,8] that corrects for the time-varying fields in a radio-frequency (rf) accelerator uses cavities that operate at the third harmonic of the main linac frequency. Two conditions must be met to approximate a dc accelerating field during pulse transit. First, the amplitude of the third harmonic is set to nine times the fundamental frequency amplitude. Second, the phase of the third harmonic is chosen to decelerate the bunch at the peak acceleration of the fundamental. The amplitude can be made flat to within 0.1% over 37° of the rf. However, the resulting two-frequency cavity will have increased phase and amplitude control complexity.

For relativistic beams, the harmonic component may be added with separate cavities, considerably reducing cavity design and control complexity. Improved accelerator performance using separated cavities for the first and third harmonic has been verified using PARMELA by Todd Smith.[7] After initial acceleration to several MeV with a long pulse (to minimize space-charge effects), the peak current is then increased using magnetic compression. The design is given in Table 1.

### 3.1.2 RF guns

The construction of the Mark III accelerator has been described in detail elsewhere.[9] The layout of the experiment is shown in Figure 5. The machine parameters are as follows: macropulse length of 2 to 5 μs, micropulse length of 2.2 ps, gun energy of 1 MeV, and a magnetic compression of 10 from the alpha magnet. The alpha magnet is also a momentum filter and limits the electron energy spread to less than 0.5%.

The electron source in the Mark III is a $LaB_6$ cathode. The cathode produces electrons by thermal emission. However, because the electrons are emitted at all phases of the rf, many of the electrons are accelerated at the wrong phases for matching into the main linac. The electrons accelerated at the wrong phases degrade the cathode because of the electron back bombardment. The peak current was 33 A. The gun x by y emittance was approximately 2 by 4 π mm-mrad, respectively.

Table 1 Injector designed by T. Smith for the Stanford High-Energy Physics Laboratory's Superconducting Accelerator[7]. Calculated emittance using PARMELA is 5 $\pi$·mm·mrad. The blocks indicate the beamline structure, and under each block are the beam energy, pulse length, and peak current.

| 300 keV ELECTRON SOURCE | HARMONIC BUNCHER | HARMONIC ACCELER- ATOR | MAGNETIC BUNCHER |
|---|---|---|---|
| 300 keV | 300 keV | 2.5 MeV | 2.5 MeV |
| 333 ps | 100 ps | 100 ps | 5 ps |
| 3 A | 10 A | 10 A | 200 A |

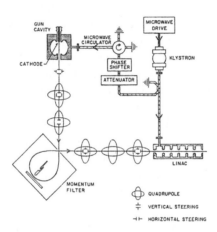

Figure 5. Schematic of the experiment showing microwave feed system and the path of the electrons from the laser-switched thermionic gun to the Mark III accelerator.

## 3.2 Photocathodes

Photoemitters can produce very high charge densities. For example, $Cs_3Sb$ photoemitters have effective temperature of 0.2 eV or greater.[10] The electron thermal temperature is not simply the difference between the incident photon energy and the semiconductor band gap (a difference of 0.7 eV for $Cs_3Sb$) because of phonon scattering in the semiconductor crystal lattice. This cathode is capable of delivering[19] over 600 A/cm², giving a minimum emittance of 0.07 $\pi$ mm-mrad for a 1 cm² cathode. However, the source emittance is a small fraction of the final emittance of the beam. Instead, the acceleration process and transport through a beamline can increase the beam emittance by over an order of magnitude.

A design that effectively utilizes photocathodes is called a photoinjector and is described in Section 4.

Photocathodes can be divided into two classes based on quantum efficiency (QE): low QE and high QE. Low QE cathodes are characterized by having reduced vacuum requirements and are relatively easy to produce. The low QE cathodes fall into two groups, metals and thermionic emitters.

### 3.2.1 Low quantum efficiency

Many different metals have been considered for photoinjector cathodes. Copper and magnesium[11] are the most common choices. Other metals that have been considered are: Al, Au, stainless steel, Sm, Y, W, Zn, Au, Mo, Ta, Pd, Zr, Ba, Na, and Ca.[12,13,14] Measurements of quantum efficiency vary considerably among individual researchers. This variation can in part be attributed to differences in samples, preparation techniques, and contamination before and during measurements. Also the UCLA group has reported non-uniform emission occurring after use in a photoinjector.[15] Overall, the measured quantum efficiency of metals varies from less than $10^{-8}$ to 3 x $10^{-3}$ near a wavelength of 250 nm.

The thermionic emitters, $LaB_6$[16] and BaO, have also been used as cathodes, both heated and unheated. Again, the measured quantum efficiencies are dependent on many factors and varies among laboratories. Quantum efficiencies of greater than $10^{-4}$ have been measured. The temporal response in the picosecond regime for these cathodes has not been measured.

Another metal cathode being used in the ATF at the Kharkov Institute is a pressed pellet of BaNi. They quote a QE of 1.7 x $10^{-3}$.[17]

### 3.2.2 High quantum efficiency

High QE photocathodes usually require a good vacuum and have a more sophisticated fabrication procedure. These types of cathodes can be subdivided into three categories: multialkali, crystal-like, and GaAs.

The $Cs_3Sb$ multialkali cathode was the first cathode used in a photoinjector. Since then, a large number of other multialkali cathodes have been used, such as $CsK_2Sb$, AgO:Cs, CsNaKSb, $K_3Sb$, and $NaK_2Sb$.

Multialkali cathodes have a significant advantage over metal cathodes. These cathodes have QE's over 2% at 532 nm, making the drive laser requirements less stringent. Unfortunately, since they rely on a cesium barrier to reduce the surface work function to near zero, they tend to be very susceptible to contamination and require $10^{-10}$ torr vacuum systems. Because of contamination issues, these cathodes have limited lifetimes.

The crystal-like cathodes, $Cs_2Te$, CsI, $K_2Te$ all require laser wavelengths shorter than 250 nm for quantum efficiencies over 2%. Their advantage is that they can survive in $10^{-8}$ vacuum systems.[18,19] Also, these cathodes can be rejuvenated by heating to 150 C and then reused.

Finally, GaAs has been used for many years as a polarized electron source. KEK plans to use this cathode in a specially cleaned photoinjector that exhibits almost no change in impurities and background pressure with and without rf power.[20]

There is a wide variety of photocathodes to choose from based on the system requirements. The photocathode, though difficult, is no longer a major impediment to using this technology.

## 4. PHOTOINJECTORS

A schematic of a photoinjector is shown Figure 6. From the first use of a photoinjector in 1985,[21] many different systems have been designed to meet the needs of very different applications. The applications include high-average-current electron beams, high-brightness sources for free-electron lasers and colliders, high pulse charges for wakefield accelerators, high-duty factor picosecond high-energy x-ray pulses, and picosecond soft x-rays by Compton back-scattering. The advantage of this source for Compton back-scattering is that the drive laser for the photocathode can be used as the scattering laser. Using the drive laser provides sub-picosecond synchronization of the electron pulse and the laser pulse.

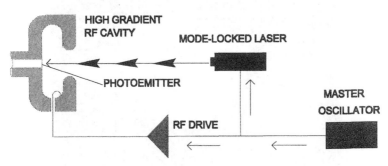

Figure 6. Basic components of a photoinjector are a laser, a photocathode, an rf source, and an rf cavity

Photoinjectors[21] have several unique characteristics. A high gradient rf cavity is used to supply the accelerating field. The high-gradient not only reduces space charge effects, but the gradient also enables laminar flow from the cathode through the accelerator to the beamline. Since the electron beam does not undergo transverse or axial mixing, a large fraction of the emittance growth due to space charge can be corrected by a technique called emittance compensation,[22] described in Section 4.6. The high gradient also allows the extraction of high charge for closely spaced pulses enabling the production of high-average currents.[23]

Since the electron source is a photocathode illuminated with a laser, the machine designer has complete control over the spatial and temporal characteristics of the electron emission process. Figure 7 is a demonstration of the spatial control of the electron beam by placing a mask in the laser beam that illuminates the cathode. Also, the gun can directly produce very short electron pulses limited only by the gun gradient and charge in the pulse. For instance, 1 nC from a cathode with a surface gradient of 30 MV/m will have a 6 ps pulse length.

### 4.1 First Photoinjector Experiment

The motivation for the first photoinjector experiment (Figure 8), was the need for an electron source that has an rms emittance of less than 10 $\pi$ mm-mrad and the capability of generating greater than 1 A average current. The ability of a photoinjector to achieve that level of performance was demonstrated in this first experiment (Figure 9).

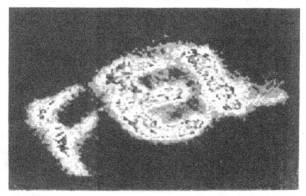

Figure 7. Electron beam image taken by a camera viewing an optical transition radiation screen 7 meters downstream of the cathode. The electron beam energy was 17 MeV. The photocathode was illuminated by a laser with a mask placed in the beam. The letters "FEL" were cut into the mask.

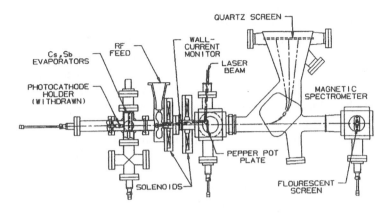

Figure 8. The first photoinjector experiment demonstrated an emittance of less than 8 π mm-mrad at 10 nC, a maximum 27 nC per 53 ps long micropulse, and 2.9 A average current for a 6 μs long macropulse. The current density was estimated to be 600 A/cm$^2$ from a Cs$_3$Sb cathode.

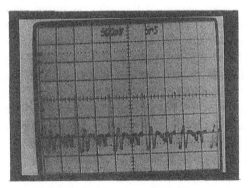

Figure 9. The first photoinjector experiment generated 2.9 A average current for 10 μs at 1 MeV, with peak micropulse currents of 390 A and micropulse charge of 27 nC. The horizontal scale is 5 ns/div. The vertical scale is 13 nC/div.

### 4.1.1 Experimental Design

The photocathodes for that experiment were fabricated in a preparation chamber vacuum coupled to the rf linac. Following fabrication in the preparation chamber, the photocathode is inserted into the rf cavity.

When the quantum efficiency of the photocathode decreases below some arbitrary minimum value, the substrate was pulled back and heat cleaned at 400°C. A new photocathode was then fabricated over the existing substrate without opening the UHV system.

The photocathode was illuminated with a frequency-doubled Nd:YAG laser. The laser was mode locked at 108.33 MHz, the twelfth subharmonic of 1300 MHz. The mode-locking crystal was driven by the same master oscillator that drove the 1300-MHz rf klystron and was therefore phase-locked to the rf. The laser generated 100-ps pulses at 1.06 μm that, after frequency doubling to 532 nm, became 70-ps-long pulses. The average power available at 532 nm was approximately 250 kW over 10 μs.

### 4.1.2 Experimental Measurements of Emittance

The experimental parameters for the emittance measurements were 11 nC (200-A peak), 70-ps Gaussian temporal width, less than 0.4-cm beam radius at the cathode (was not accurately measured at the time of the experiment and only the upper bound is known), 1.0-MeV beam energy, and a solenoid field of 1.8 kG. The measured emittance was 10 π mm-mrad. The measured emittance did not agree with a PIC simulation (which gave greater than 35 π mm-mrad) of the experiment. This disagreement led to a detailed examination of the gun, beamline, and the pepper-pot emittance diagnostic using PARMELA and MASK[24] simulations.

The experimental and simulated electron-beam diameter at the pepper pot and the diameters of the beamlets produced by the pepper pot at the second quartz screen are in close agreement, confirming the accuracy of the simulations. The emittance of the electron beam for that experiment, with 10 nC per bunch, was calculated from the simulations to be 30 π mm-mrad for 100% of the beam. Simulations[24] show that, if the beam is clipped in time and left with 80% of the original charge, then the emittance of the remaining beam was calculated to be 10 π mm-mrad in agreement with the experimental results. The results of the MASK calculations are shown in Figure 10.

Since most applications of bright electron beams depend upon only the bright central core of the electron bunch, the low emittance given by neglecting the temporal tails of the distribution is an accurate estimate of the electron beam performance. More importantly, the accuracy of the simulation codes was verified for the future linac design.

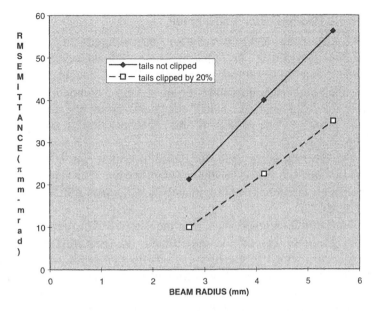

Figure 10. The beam emittances from MASK simulations (performed by Bill Herrmannsfeldt of SLAC) are within the experimental error in beam radius if the temporal tails of the Gaussian pulse are not included. The two curves show the difference in emittance gained by excluding a small fraction of the charge at the front and tail of the pulse.

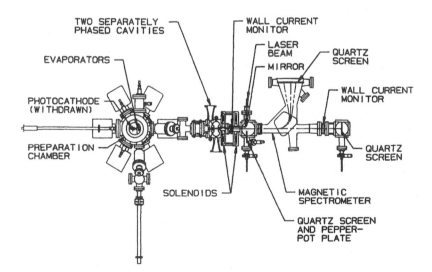

Figure 11. Two-cavity experiment showing gun, beam transport, and diagnostics.

## 4.2  Second Photoinjector Experiment

A second experiment using two-cavities, shown Figure 11, followed the single-cavity experiment. The linac had two 1300-MHz rf cavities with independent amplitude and phase controls. Both rf cavities had loops to measure the phases and amplitudes of the rf fields. Following the second cell were the diagnostics for bunch charge, beam energy, emittance, and temporal profile. The details of the rf cavity design are presented elsewhere.[25]

The electron energy gain for typical operation was 0.9 MeV in the first cavity and 1.8 MeV in the second cavity. This corresponds to operating both cavities at approximately 2 Kilpatrick (58 MeV/m peak surface field).

The aforementioned laser for the single-cavity experiment was modified for the two-cavity experiment. A Spectra-Physics pulse compressor was added to the optical train for generation of 4- to 20-ps pulses. The laser pulse length was limited by the gain bandwidth of the Nd:YAG amplifiers to approximately 16 ps. The maximum charge extracted for this pulse was 13.2 nC from 1 cm² of photocathode surface. This gives 820 A/cm² of current density at the cathode. However, PARMELA simulations predict that a 16-ps electron pulse increases to 22 ps on passage through the first cavity, giving a peak current after the first cavity of 600 A.

## 4.3  Mark III Photoinjector

The microwave rf gun for the Mark III accelerator was also run in a photoinjector mode. The current emission from the cathode was limited by average-power heating; therefore, by using a laser to limit the emission to the correct rf phase, higher peak currents can be obtained.[26] Also, the lifetime of the cathode is expected to be longer. In this mode, the $LaB_6$ cathode was operated just below its normal emission temperature, and a laser was used to pulse the cathode. Operation with the laser resulted in an increase in peak current from 33 A to 75 A with no observable increase in beam emittance.

## 4.4  Photoinjector Lasers

The key to the stability of a photoinjector is the drive laser. The advantage of using a laser is that the cathode can be illuminated with any temporal and spatial profile required to optimize the gun performance. Lasers have excellent temporal stability, with almost all of the present systems having temporal jitter less than a picosecond. Also, if only single pulses are required, a laser can generate very large energy per pulse (a Table-Top Terawatt laser (can generate 1 J in less than 1 ps).

The remaining difficulty in the laser systems is the macropulse to macropulse amplitude stability. Achieving less than 10% amplitude stability is very difficult with present systems. The technology exists to achieve less than 1% stability, but not the resources. Present laser stability measurements[27] are less than 0.5 ps and less than 10% amplitude variations.

Lasers can generate high peak energy in short pulses easier than long (many microseconds) pulses. It follows that for long pulse trains a minimum QE of 0.5% is required.

One other issue that can be critical to stable operation is pointing stability. Since the laser defines the spatial profile of the emission, the laser must be stably pointed at the cathode. For example, the large solenoid around the gun region of the Advanced Free-Electron Laser[28] acts to amplify small transverse spatial variations of the cathode position. This amplification occurs because of the long distance from the large solenoid to the first focusing element (2.5-m lever arm). Even though the cathode diameter is 8 mm, a shift of 100 microns in the centroid will image to a 25-micron shift in the middle of the wiggler. A 25-micron transverse shift will significantly degrade the performance of an FEL.

## 4.5 Present Photoinjector Designs

Photoinjectors routinely generate greater than 500 A/cm$^2$. For most systems this current density is limited only by the field gradient on the cathode or the laser intensity.

Research is still proceeding on high-average current machines at Boeing Defense and Space,[29] and at Bruyeres-le-Chatel.[30] The first demonstration of a high-average current using a photoinjector was on the Boeing accelerator. This 25% duty factor machine has demonstrated an average current of 32 mA at 5 MeV, giving an average beam power of 160 kW. The macropulse average current was 0.13 A. The beam emittance was 5 to 10 $\pi$ mm-mrad for 1 to 7 nC pulse charge. An example of the machine located at Bruyeres-le-Chatel is shown in Figure 12.

Many designs are based on the work done at Brookhaven National Laboratory at 2856 MHz.[31] A schematic of one of their guns is shown in Figure 13. The Brookhaven type of gun is being used for advanced accelerator studies, free-electron lasers, and linear collider injectors. One of the advantages of operating near 3 GHz is the higher cathode surface electric fields that can be obtained relative to operating at lower frequencies.

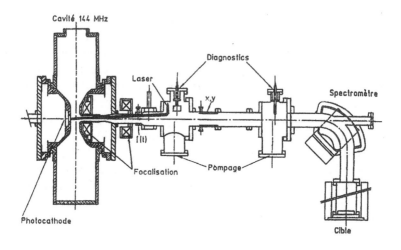

Figure 12. Photoinjector at LEL-HF Bruyeres-le-Chatel. RF cell produces a 2.0 MeV beam at 5 nC with a 20 to 50 ps pulse length. They have measured 4 π mm-mrad at 1 nC.

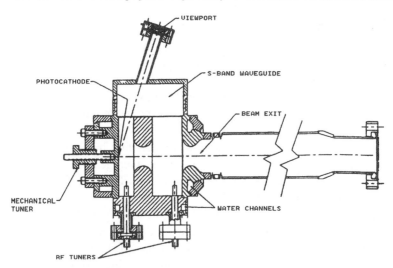

Figure 13. Brookhaven's 2856 MHz photoinjector operates at 3 MeV and has produced 4 π mm-mrad at 1 nC with a cathode field of 70 MV/m. The gun has generated 4.5 MeV beams. This photoinjector uses two cells operating in a π-mode configuration with a single radio-frequency feed. Copper or magnesium are the typical cathode materials.

A new photoinjector operating at 17 GHz has been constructed at the Massachusetts Institute of Technology (MIT). This gun has 1-1/2 cells with peak surface fields of 250 MV/m and a peak cathode surface field of 200 MV/m. The rf source is a gyro-amplifier developed at MIT. [32]

Photoinjectors have generated micropulse charges between 1 and 30 nC. However, Argonne National Laboratory has generated greater than 60 nC per micropulse.[33]

Electron pulse lengths are limited by space charge effects in the first few centimeters in front of the cathode. Typically, less than 10 ps pulses is generated for 1 nC of charge in a micropulse (in the AFEL 6 ps for 1 nC)

The measured electron beam rms emittance varies, depending on the machine design, between 1 and 5 $\pi$ mm-mrad for 1 nC in a micropulse. Some newer designs give less than 1 $\pi$ mm-mrad for 1 nC by scraping the wings of the distributions.[34]

## 4.6 Emittance Compensation

Emittance compensation[35] is a technique for decreasing the growth in emittance that occurs in a low-energy charged particle beam with large space charge forces. The technique uses a focusing element (a solenoid or quadrupole) to correct for the distortions in phase space that occurs when a beam expands due to the space charge forces. A representative configuration is shown in Figure 14.

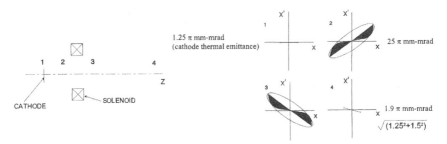

Figure 14. At the cathode, the electron bunch emittance is determined by the cathode's thermal emittance (position 1). As an electron bunch leaves the cathode, the bunch expands radially due to radial space charge forces. Since the space charge force acts continuously on the bunch, no single discrete lens can compensate for the distortion of the distribution in phase space (position 2). However, if a simple lens can be used to focus the bunch (position 3), then, the same types of forces that acted on the bunch during expansion are present while the bunch is focused (position 4). The figure also shows the residual emittance (1.5 $\pi$ mm-mrad) after emittance compensation is used. A thermal emittance of 1.25 $\pi$ mm-mrad is the cathode emission temperature of $Cs_3Sb$.

A fully non-linear model for emittance compensation cannot be solved analytically. However, simplified models can be solved. To demonstrate the basic principles, our model has the following assumptions. A beam with linear radial electric fields. The beam has non-relativistic velocities perpendicular to propagation. The beam radius does not change significantly and therefore the force is approximately constant with respect to the radius and time. The beam longitudinal velocity is approximately the speed of light. With these assumptions the radial electric field, $E(r,\zeta,t)$, is

$$E(r,\zeta,t) = -m*k(\zeta)*\rho/e, \qquad (4.1)$$

where m is the electron mass, $k(\zeta)$ is the force constant that can vary with longitudinal position, $\zeta$ is the relative longitudinal coordinate, $\rho$ is the relative radial position, and e is the electron charge. Figure 15 shows the relationship between the relative and absolute coordinates.

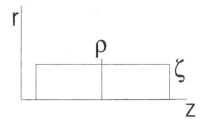

Figure 15. A plot of the relative coordinates $\rho$ and $\zeta$ with respect to radius, r, and longitudinal position, z.

The radial force, $F(r,\zeta,t)$, is

$$F(r,\zeta,t) = m*d^2r/dt^2 = -e* E(\rho,\zeta,t) = m*k(\zeta)*\rho \qquad (4.2)$$

Then the radius is

$$r = k(\zeta)*\rho*(t-t_0)^2/2 + dr/dt_{(t=t_0)}*(t-t_0) + r_0, \qquad (4.3)$$

and the radial velocity is

$$dr/dt = k(\zeta)*\rho*(t - t_0) + dr/dt_{(t=t_0)}. \quad (4.4)$$

Setting $z = c*t$, $dr/dt = (dr/dz)*(dz/dt) = r'*c$ and $r'_0 = dr/dt_{(t=0)}/c$, then

$$r = k(\zeta) \bullet \rho \bullet (z-z_0)^2/2/c^2 + r'_0 \bullet (z-z_0) + r_0 , \ r' = k(\zeta) \bullet \rho \bullet (z-z_0)/c^2 + r'_0. \quad (4.5)$$

In phase space a useful relationship is the ratio between r and r'. The closer this ratio is to a constant, the straighter the line in phase space and the lower the emittance. Solving for the ratio gives

$$r'/r = [k(\zeta) \bullet \rho \bullet (z-z_0)/c^2 + r'_0]/[ k(\zeta) \bullet \rho \bullet (z-z_0)^2/2/c^2 + r'_0 \bullet (z-z_0) + r_0]. \quad (4.6)$$

Assuming the initial conditions $z_0 = 0$, $r = r_0$ and $r'_0 = 0$, then

$$r'/r = [k(\zeta) \bullet z/c^2 ]/[k(\zeta) \bullet z^2/2/c^2 + r_0/\rho] \sim k(\zeta) \bullet z/[c^2 \bullet r_0/\rho], \quad (4.7)$$

therefore r'/r is approximately proportional to $\rho$. This gives a curved line in phase space for each location $\zeta$ as shown in Figure 16. By focusing this beam with a simple linear lens with a focal length, f, given by

$$f = (z_d - z_1)^2/(2z_d) \quad (4.8)$$

and located at position $z_1$, it is possible to fully compensate at position $z_d$ for the curvature in phase space.

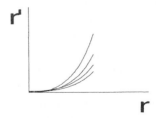

Figure 16. Phase space in r versus r' for a beam that has expanded under a linear space charge force. Each curved line represents a differing force constant $k(\zeta)$.

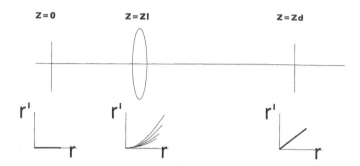

Figure 17. Plots of a beam with different space charge forces in longitudinal slices being focused with a lens at position $z_l$. At position $z_d$ the beam slices overlap and are straight lines, giving a zero emittance beam.

The ratio of r/r' is then

$$r'/r = 2z_d/(z_d^2-z_l^2). \qquad\qquad (4.9)$$

Since r'/r is independent of $k(\zeta)$ and $\rho$, we get a superposition of straight lines in phase space for all $k(\zeta)$, as shown in Figure 17.

The effect of space charge on a finite length pulse for larger variations in radius is not analytically soluble. However, the same focusing technique can correct for emittance growth. A first order expansion of the equations was solved exactly by Carlsten.[36]

## 5. ADVANCED FREE-ELECTRON LASER

A new accelerator design that produces a very bright electron beam in a compact form has been developed through the Advanced Free-Electron Laser Initiative[37] (AFELI) at Los Alamos National Laboratory. The goal of AFELI was to build a second-generation free-electron laser (FEL). This FEL was designed to be suitable for a wide range of industrial, medical, and research applications. State-of-the-art components were incorporated so that the FEL system is compact, robust, and user friendly.

The accelerator design incorporates the experience from the initial photoinjector experiments described earlier in his paper and the later accelerator experiments at APEX (APLE Prototype EXperiment).[38] The design simulations were performed using a modified version of the code PARMELA.

The emittance is calculated in two ways. The "full" emittance is calculated by using the entire micropulse in time and space. The "slice" emittance is calculated by dividing a micropulse into slices in time equal to a slippage length. To ensure enough particles are in a slice to give reasonable statistics, the smallest time slice is limited to 1% of the total pulse length. We calculate the slice emittance because the electrons do not generate gain over the entire pulse, but only for the middle portion (in time) of the pulse. The individual slices can have different divergence, and so only a few of the slices may be properly matched and not the entire pulse. If temporal mixing occurs, the use of slice emittance is invalid and the full rms emittance must be used. To minimize mixing and to preserve beam brightness, great care must be given to proper beam-line design.[39]

## 5.1 Accelerator Construction

The design goals for the AFEL accelerator are to maximize beam brightness, develop a simple design, and operate at the relatively high duty factor of 0.1%. The design point is greater than 2 nC charge per micropulse and an effective emittance of less than 10 $\pi$ mm-mrad. Simple design is accomplished by using a single radio-frequency feed to drive the entire accelerator structure. The $\pi/2$ mode, 10-1/2 cell accelerator (Figure 18) has the following features: 20-MeV output energy, average cavity gradients of 22 MeV/m, 10-Hz repetition rate, 20-$\mu$s long macropulses, 8- to 20-ps long electron micropulses. The accelerator can be operated at liquid-nitrogen temperatures. The accelerator is driven with a 1300-MHz, 15-MW-peak-power klystron.

At 15 MeV, the peak rf power lost to copper is 5 MW. The beam power is 7.5 MW for 0.5 A average current corresponding to 4.6 nC per micropulse at 108 MHz or 2.3 nC at 216 MHz. Assuming 20% control margin (2.5 MW above 12.5 MW), 50% of the rf power is converted into beam. Although the rf macropulse is 18 $\mu$s long, to avoid cavity transients, the electrons are turned on during the last 15 $\mu$s of the rf pulse when the cavity field is flat.

### 5.1.1 Using a Solenoid to Compensate for Emittance Growth Caused by Space Charge

The use of a solenoid to reduce emittance growth caused by space charge was discussed in Section 4.6. Thus, proper lens placement significantly reduces emittance growth. A unique solenoid design follows from the requirements of minimum emittance growth and simultaneously having the beam focused at a particular axial location. The solenoid design depends on the accelerator gradient, current density, and location of the peak magnetic field with respect to the cathode. A plot giving the magnetic field lines in the AFEL gun cell region is shown in Figure 19. The emittance numbers in Figure 14 are from a typical PARMELA run. To accurately render the solenoid field profiles, we incorporated the POISSON field maps of the solenoid directly into PARMELA.

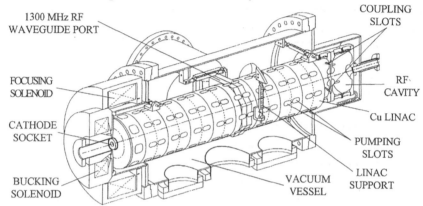

Figure 18. AFEL linac schematic.

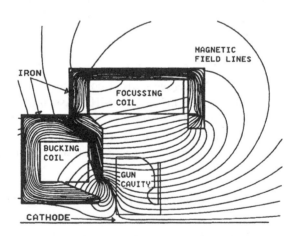

Figure 19. The AFEL linac design must have proper lens placement to minimize emittance growth.

From simulations, we computed the effect on the final emittance caused from the cathode thermal effects. As expected, the final emittance is the sum of squares of the final emittance calculated with zero cathode temperature and the finite cathode emittance. An example is shown in Figure 14.

### 5.1.2 Minimizing Perturbations caused by Accelerator Coupling Slots

The APEX accelerator[38] was the first two-cell-photoinjector with more than two cells. The standing-wave, 1300-MHz, π-mode accelerator is designed with on-axis coupling slots. The initial PARMELA simulations gave a symmetrical beam at the accelerator exit (Figure 20). In the APEX experiment, however, the accelerator produced elliptical beams. By incorporating MAFIA field maps of the coupling slots into PARMELA, we found that the coupling slots produced a quadrupole lens in every accelerator cell. A sample output plot of the APEX photoinjector from the modified PARMELA is shown in Figure 21.

Several possible configurations of on-axis coupling are shown in Figure 22. A single slot produces a dipole lens. Two slots produce a quadrupole lens. Four slots produces an octupole lens. Each accelerator cell (except the cells at the accelerator ends) has two sets of coupling slots. The two-coupling-slot configuration gives a quadrupole lens at the entrance and exit of the accelerator cell. The orientation of the slots will determine whether the quadrupole lenses add or subtract in focusing power. In the APEX arrangement (type H) the fields at each cell end are additive, giving a net quadrupole lens. In a type T arrangement the fields at each cell end subtract, giving a net effect close to zero.

The cancellation of the quadrupole effects in a type T arrangement is nearly zero only for a highly relativistic beam. In the first few cells, where the beam is still not highly relativistic, a net quadrupole lens still exists. Also, the large solenoid around the gun cell causes the beam to rotate azimuthially. If a quadrupole field overlaps the solenoid field, the x and y phase spaces will mix, causing an irreversible growth in emittance.

The coupling-slot design for the AFEL accelerator uses a four-coupling-slot arrangement for the first two cells and a type T configuration for the remaining accelerator cells. Because the four-slot arrangement has no quadrupole component, the first two cells produce no beam asymmetry. After the beam exits the first two cells, the beam is highly relativistic and the type T coupling gives a very small net quadrupole focusing. This configuration is shown in Figure 23.

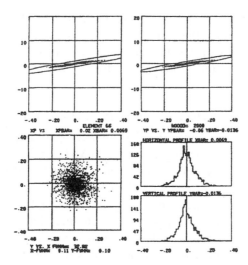

Figure 20. PARMELA simulation result neglecting the effects of coupling fields. The upper left picture is the x' versus x phase space, the upper right is the y' versus y phase space, the lower left is the beam spot plotted as y versus x, and the lower are the horizontal and vertical profiles. X and y are in centimeters, and x' and y' are in milliradians.

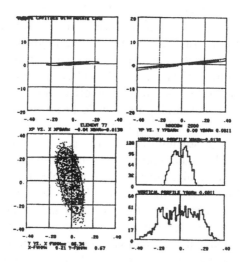

Figure 21. PARMELA simulation result with the type H coupling-cell configuration.

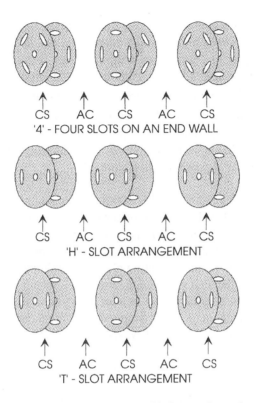

CS    AC    CS    AC    CS
'4' - FOUR SLOTS ON AN END WALL

CS    AC    CS    AC    CS
'H' - SLOT ARRANGEMENT

CS    AC    CS    AC    CS
'T' - SLOT ARRANGEMENT

Figure 22.  Possible coupling-slot arrangements with four- and two-slot coupling. CS is a coupling cell.  AC is an accelerator cell.

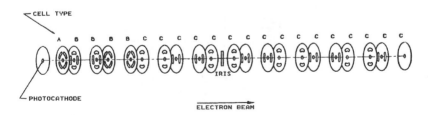

Figure 23. Schematic of the AFEL's coupling slots and accelerator cells. The A, B, and C above the cells indicate different cavity shapes.

The four-coupling-slot arrangement cannot be carried throughout the accelerator. At the high-average currents of the AFEL, beam breakup will occur because of coupling of a dipole mode from cell to cell. In the type T- and H-coupling-cell configuration, the dipole mode does not couple because the coupling slots are rotated 90° in the coupling cavity. In the 4-slot coupling-cells, the slots are rotated 45° in the coupling cavity, which very effectively couples the dipole modes.

The PARMELA simulation for the coupling slot arrangements shown in Figure 23 is presented in Figure 24.

### 5.1.3  Other Features of the AFEL Accelerator

The first cell, a half-cell, is 9 mm longer than one-half of a standard 1300-Mhz cell. A longer injection cell has two advantages. First, the exit phase of the electron bunch depends on the cell length. Since the AFEL linac has a single rf feed, the proper operating phase to minimize energy spread was met by adjusting the first cell length. Second, a longer first cell increases the electron-beam energy at the exit of the first cell. This reduces the space-charge effects and helps improve the final emittance. The exit energy from the first cell is 1.5 MeV instead of 1.0 MeV for a regular half-cell.

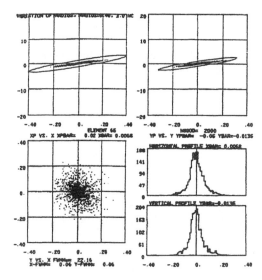

Figure 24.  PARMELA simulation result for the AFEL accelerator, including all of the coupling slot effects..

Other engineering features of the AFEL accelerator are the capability of operation at 77K; UHV design; and high-Q, high-gradient accelerator cells.

### 5.1.4 Beam Dependencies

This type of accelerator is unique in that the electron-beam distribution does not mix longitudinally. With no mixing, the rms emittance calculation for the full pulse underestimates the FEL performance. Figure 25 shows the x- and y-slice emittance during a micropulse for a Gaussian and a square pulse. Except for statistical noise caused by the limited number of particles in the simulation, the slice emittance is time independent during the micropulse. However, the emittance of the full pulse is significantly larger. The larger full-pulse emittance is caused from the variation in divergence throughout the micropulse (see upper graph in Figure 26). Two factors help determine FEL performance: first, the local beam conditions in the micropulse (since the slippage length is a small fraction of the entire pulse length); second, the ability to match into the gain profile of the wiggler. Figure 26 shows the beam conditions that affect FEL performance. The upper two graphs are the beam divergence and the particle density as a function of time. The lower graph is a calculation of $\Delta v$[40] (gain width for a sample wiggler) as a function of time. The three graphs show that most of the electrons are in the gain width of the wiggler for the middle portion of a micropulse. The beginning and end of the micropulse are not matched into the wiggler, but the fraction of the electrons in the temporal wings is small. Again, this type of analysis is not correct if the beam mixes longitudinally.

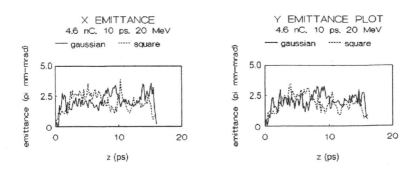

Figure 25. Plots of the slice emittance during a micropulse. The plots give results for pulses that are either Gaussian or square in time. The beam had a square spatial profile on the cathode.

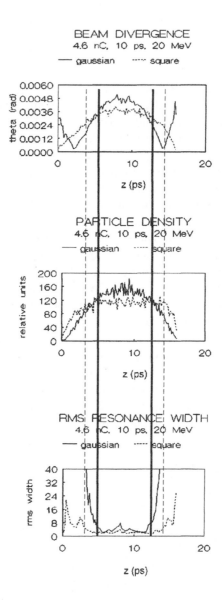

Figure 26. The upper plot shows the beam divergence during the pulse. The middle plot shows the charge density during the pulse. The bottom plot shows how well matched the pulse is to the gain profile of the wiggler. The solid (dotted) lines through all three plots represent the fraction of a Gaussian (square) beam that is well matched to the AFEL wiggler.

The AFEL is designed to minimize components and distances and to increase reliability and ease of use. However, the performance of the FEL design does depend strongly on a few parameters. The parameters that must be tightly controlled are: the radius of the cathode, the magnitude of the solenoid field around the cathode region, the centering of the cathode relative to the center of the magnetic field, the accelerator phase, and the magnitude of the accelerator fields.

### 5.1.5 Engineering Issues

The accelerator is typically run between 13 to 19 MeV. At a beam energy of 18 MeV, the first cell gradient is 23 MV/m and the remaining accelerator cells' gradients are 20 MV/m.

Field emission from the cathode depends on the surface properties and gradient present in the cathode cell. Designs of high-duty-factor machines or experiments where the background charge levels are important must include the effect of the field emission. The field emission can occur either directly from or near the photocathode. Thus the maximum cell gradient might be limited, impacting the beam emittance. High-QE cathodes seem to exhibit more field emission than the low-QE cathodes. The surface of the cathode can also impact the field emission. Cathodes in which the machining grooves can be seen exhibit much high field emission than cathodes given a mirror polish.

The spatial profile of the photocathode laser and photocathode QE is important. In general, spatial profiles in which the electron density decreases with radius, as opposed to radially uniform or slightly increasing with radius, produce a greater emittance.

To preserve a bright electron beam throughout the beam-line requires detailed simulations. The AFEL beam line was designed[4] and the parametric sensitivities measured[41] using PARMELA.

### 5.2 Laser and Photocathode

The photoinjector front-end consists of a drive laser and a $Cs_2Te$ photocathode. The drive laser starts with a Nd:YLF oscillator mode-locked at 108.33 MHz (12th subharmonic of the accelerator frequency). A transverse Pockel's cell switches out variable-length, programmable macropulses to be amplified in a double-pass amplifier. Frequency doubling in a lithium triborate crystal converts 50% of the 1.05 μm infrared light to 527-nm green light. Following the doubling crystal is a quadrupling crystal, generating 263-nm light. A typical 15-μs macropulse consists of 1,600 micropulses, each approximately 10 ps long. Space charge effects increase the electron pulse width depending on the beam energy and the charge density. For 15-MeV beams, an illuminated area of 0.4 $cm^2$ and a micropulse charge of 3 nC, the calculated electron pulse width is 15 ps.

The photocathodes exhibit a typical quantum efficiency (QE) of 10% after they are fabricated. Due to poisoning during the transfer process, their QE is reduced to 5% in the linac. A typical cathode's useful lifetime has not been systematically measured. To extend beam time and to provide flexibility in trying new cathode materials, the photocathodes are prepared six at a time (a "6-pack") in a separate preparation chamber. The 6-pack is then transported under vacuum to the accelerator via an actuator/inserter mechanism. The time required between preparation of new six-packs is greater than a month.

### 5.3  Beamline Construction

The Advanced FEL (Figure 27) with a 1-cm-period wiggler has been in operation since early 1993 in the spectral region between 4.5 and 6 μm, limited by the bandwidth of multilayer dielectric optics.[42] With metal mirrors, the Advanced FEL can be tuned from 4 to 12 μm. A broader tuning range (3 to 20 μm) can be achieved by lasing at fundamental and harmonic frequencies with a high-field, 2-cm-period wiggler.

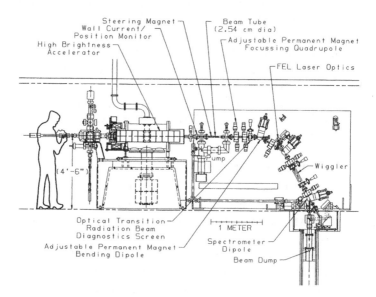

Figure 27. Layout of the AFEL beamline showing cathode insertion mechanism, accelerator, electron optics, and FEL optics.

### 5.3.1 Electron Beam-line

The beam-line consists of three 30-degree bends. The first bend allows direct visual access to the cathode. Thus the drive laser beam is line-of sight with the cathode. The second bend deflects the electron beam into the FEL optical axis. The third bend deflects the beam into an electron energy spectrometer. The electron beam then goes into a carbon-block beam-dump. The first two bends are made achromatic by a single quadrupole that refocuses the beam in the dispersion plane. The beam is matched into the two bends by setting the first two quadrupole doublets so that the beam is focused in x-plane (the bend plane) in the center of the dipoles and in y-plane at the quadrupole singlet. For compactness and reliability, all focusing quadrupoles and bending dipoles are made out of permanent magnets. Detailed designs of the variable-field permanent magnet quadrupoles and dipoles have been reported.[43] The use of permanent magnet components offers two advantages: a) once the electron beam is aligned through the wiggler, lasing can be reproduced every day with only minor adjustments of the beam optics, and b) the permanent magnet optics do not need power or cooling, thus simplifying the design.

To transport the beam through the beamline around the bends and through a 2.8-mm-id wiggler tube, we relied on TRACE 3D to set the quadrupoles to the approximately correct values. The beam profiles were then measured using optical transition radiation (OTR) screens and matched to those calculated by TRACE 3D by adjusting the quadrupole field. Wall-current, beam-position monitors (BPM) were used to monitor beam transport through the center of the beamline and also to measure relative beam current. The BPM performance has been reported elsewhere.[44] The BPM beam current was monitored to ensure 100% beam transmission through the wiggler tube.

The electron beam energy was measured with the first magnetic dipole using beam position monitors before and after the first bend to determine the beam centroid position. The error in electron beam energy measurements is approximately ± 2%. A high-resolution energy spectrometer at the third bend was used to measure beam energy spread and fluctuation to a precision of ± 0.025%. A typical energy spread, integrated over the 12-μs macropulse and thus including energy slew, is 0.5%. The energy fluctuation from macropulse to macropulse is ± 0.25%. The actual micropulse energy spread has not been measured, but the FWHM energy spread is calculated to be 0.25%.

## 5.3.2 Wiggler

For oscillation in the 4 to 10 µm region, a 24-cm untapered linear wiggler with 1-cm periods is used. Each period consists of two pairs of samarium cobalt magnets arranged with the magnetization oriented along the beam axis. The wiggler is shown in Figure 28.

The measured on-axis field was 0.42 T with a gap of 3 mm. Due to the orientation of the magnets, the third harmonic content reduces the fundamental field by 0.02 T. The fundamental field is thus 0.4 T, yielding an rms wiggler parameter $a_w$ of 0.266. Before assembly on the beamline, the wiggler was tested via the taut wire technique.[45] After the wiggler was mounted on the optical table, the area around the wiggler was found to have a constant field of approximately 4 gauss. A small correction electromagnet was used to cancel out this residual magnetic field.

## 5.3.3 Resonator optics

The Advanced FEL optical resonator consists of two concave mirrors mounted in vacuum with the wiggler at the center of the resonator. The 1.3836-m mirror separation and 0.70-m radius of curvature give a resonator parameter (g = 1 - L/R) of -0.9766 and Rayleigh range of 7.5 cm

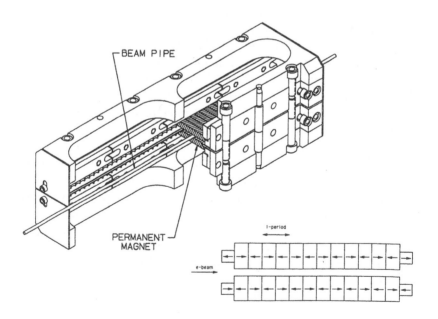

Figure 28. AFEL permanent magnet wiggler with a 1 cm period, 24 periods, and a 0.3 cm gap. The rms $a_w$ is 0.27.

(about one-third of wiggler length). At 6 μm, the calculated lowest order mode size is 378 μm at the center of the wiggler and 753 μm at the end of the 26.3 cm-long wiggler tube. The 2.8-mm-id tube introduces an estimated loss of 0.2% for the lowest order transverse mode. However, this vignetting loss can be substantially larger if hole coupling is used because the hole forces the optical mode to have a larger diameter throughout the resonator.

Two sets of optics have been used in the Advanced FEL resonator. The first set consists of two diamond-turned gold evaporated copper mirrors. These mirrors have an averaged reflectivity better than 99% over the mid-infrared region. However, diamond turning on concave surfaces produces grooves that scatter light and introduce an additional loss. The resonator round-trip loss with two 1% hole-coupled mirrors was measured to be ~8%. Out-coupling in metal mirrors is provided by a hole drilled in the center of the mirrors. For a hole with a radius, a, that is small compared to the empty cavity mode size, w, at the mirrors, the fraction of light out-coupled is approximately $2(a^2/w^2)$. Although metal mirrors provide broad spectral coverage, there are two problems associated with hole-coupling. First, diffraction caused by the presence of the out-coupling hole modifies the empty cavity mode in such a way that the vignetting loss at the wiggler ends increases. The vignetting loss increases as one tries to out-couple more by increasing the hole size, and so the ratio of out-coupling to total loss is relatively constant.[46] Second, alignment is difficult because the alignment HeNe beam cannot be injected through the out-coupling hole and matched into the resonator mode.

The second set of optics is $ZnSe/ThF_4$ multilayer dielectric (MLD) on ZnSe substrates. Two different MLD coatings with 99.5% and 99.0% reflectivity at 4.5 to 5.5 μm have been used. The transmission of each mirror, measured to be 0.5% to 1% over the 4.5 to 5.5 μm region, respectively, provides the out-coupling. Higher out-coupling is possible by tuning the FEL wavelength to either side of the coating reflectivity curve. The advantages of using dielectric mirrors are the low resonator loss (the round-trip cavity loss was almost entirely due to outcoupling). Another advantage is the ability to transmit the alignment HeNe laser beam through one of the mirrors. However, the MLD coatings are easily damaged in the forms of mm-size pits and cracks caused by the peak intracavity power. The coating damage limits the peak intracavity power to less than 500 MW (3 GW/cm$^2$ on the mirrors) and precludes sideband operation that would lead to much higher extraction efficiency.

## 5.3.4 FEL optical diagnostics

The FEL output was characterized with various optical diagnostics. A sensitive HgCdTe detector was used to measured the spontaneous emission, or with suitable attenuation, the coherent emission. To obtain macropulse buildup and ring-down times, a Molectron P500 pyroelectric detector with sub-nanosecond response time was used. Macropulse energy was measured, sometimes at both ends of the FEL if the outcoupling was bi-directional, with Molectron J50 and Gentec ED-200 pyroelectric energy detectors. These energy detectors are calibrated to within ± 5%. The FEL spectral characteristics were measured with an Optical Engineering spectrum analyzer. The spectrum analyzer has a 75-grooves/mm grating and a Molectron pyroelectric array at the exit focal plane.

## 5.3.5 FEL pulse energy

The highest macropulse energy was observed for an electron beam energy of 15 MeV and a peak current of 200 A was 240 mJ after correction for loss in the transport optics. The peak beam power in each micropulse was 3 GW. The mirrors used in this experiment were a 1% hole-coupled, gold-evaporated mirror with an additional loss of 3%, and an MLD mirror that has an out-coupling of 7% at 5.8 μm. The measured ring-down is 84 ± 10 ns corresponding to a total cavity loss of 11 ± 1%. The maximum output energy recorded by the Molectron J50 energy meter was 25 mJ at the 1% out-coupling mirror of the resonator through a $CaF_2$ window. The Gentec ED-200 energy meter recorded 175 mJ through an uncoated $CaF_2$ window, two Cu mirrors, and an uncoated $CaF_2$ lens. The total output energy after correcting for reflection loss was 240 mJ. Ignoring the ends of the macropulse, there are approximately 1200 micropulses in the 11-μs saturated portion of the macropulse. The energy in each micropulse was thus 200 μJ. The estimated peak optical power was 20 MW corresponding to an efficiency of converting electron beam power into light of 0.7%. The intracavity extraction efficiency is estimated to be 1%, one-half of the theoretical limit without sideband of 2% (1/2N). A sample 8-μs long macropulse is shown in Figure 29.

## 5.4 Schedule

The AFEL was a five year program funded with Los Alamos institutional funds. Table 2 is a list of the accelerator and beamline commissioning activities. Table 3 is a list of the free-electron laser commissioning activities.

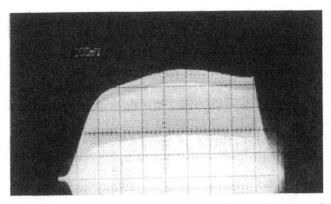

Figure 29. The AFEL optical output. The horizontal divisions are 1 μs/div and the vertical scale is output power in relative units.

Table 2. A history of AFEL accelerator commissioning activities

| Mar 1990 - Jan 1991 | Simulations |
|---|---|
| Jan 1991 - Jan 1992 | Design and Off-site Fabrication |
| Jan 1992 - March 1992 | Linac Commissioning |
| July 1992 | First Electron Beam |
| Sep 1992 - Nov 1992 | Beam to Spectrometer |
| January 1993 | 100% Beam Transport |

Table 3. A history of the AFEL laser commissioning activities

| February 1993 | Spontaneous Emission at 5.2 mm (~0.1 mW) |
|---|---|
| April 1993 | Lasing at 4.7 mm (7 mW) |
| October 1993 | 34 mW |
| December 1993 | 1.6 Watts |

## 5.5  Experimental Results

Emittance measurements in a photoinjector are complicated by one of the photoinjector advantages. Because of the rapid acceleration and lack of other beamline components in the gun region, the longitudinal phase space of the beam does not thermalize. As a result of the non-thermalization, different longitudinal parts of the beam propagate with their own trajectories. This complicates the analysis of the beam phase space ellipse. Commonly used techniques for measuring emittance, such as pepperpot technique or quadrupole scans, can lead to erroneous phase-space emittance measurements.

As described in Section 4.1.2, the first photoinjector experiment used a pepperpot to measure the emittance. Because of the longitudinal variations in phase-space, the emittance was underestimated by a factor of four. In this section, I will describe difficulties in using a quadrupole scan technique to determine the beam emittance.

For a thin lens, the rms unnormalized emittance, $\varepsilon_{un}$ can be calculated by fitting the beam spot size $x_s$ to the coefficients of $1/f$ in the following

$$x_s^2 = x_{min}^2 \left[1 + \{L^2 \varepsilon_{un}(1 - f_w / f) / (x_{min}^2 f_w)\}^2\right], \tag{5.1}$$

where $\varepsilon_{un}$ is the unnormalized rms emittance, $L$ is the spacing between the quadrupole and image screen, $f$ is the focal length of the quadrupole, $f_w$ is the focal length that gives the minimum spot size $x_{min}$.[47] The focal length of a quadrupole is $\beta\gamma m_e c/(leB)$, where $\beta$, $\gamma$ are the relativistic factors, $m_e$ is the mass of an electron, $c$ is the speed of light, $l$ is the quadrupole length, $e$ is the electron charge, and $B$ is the quadrupole field gradient.

For a thick or thin lens, the spot size $x_s$ can be fit using the Twiss parameters[48] with the following formula,

$$x_s^2 = \varepsilon_{un} \left[m_{12}^2 \gamma_q - 2m_{11}m_{12}\alpha_q + m_{11}^2 \beta_q\right], \tag{5.2}$$

where $\gamma_q$, $\beta_q$, and $\alpha_q$ are the Twiss parameters of the beam at the quadrupole. The coefficients of the Twiss parameters are from the Twiss parameter transfer matrix for a thick lens,

$$m_{11} = \cos(\theta) - d\theta \sin(\theta) / L$$
$$m_{12} = d\cos(\theta) + L\sin(\theta) / \theta \tag{5.3}$$
$$\theta = L\sqrt{eB / (\beta\gamma mc)}$$

where $d$ is the quadruple axial length and $L$ is the spacing from the end of the quadrupole to the image screen. Using the identity $\beta_q\gamma_q - \alpha_q^2 = 1$, the rms unnormalized emittance can be calculated from the coefficients of the fit.

The experimental data and the fit using the Twiss parameters are shown in Figure 30. The thin lens fitting procedure was also used on the experimental data. The data and the fit were within 10%. The rms emittance as calculated from the fit for either the data spot sizes or for the PARMELA spot sizes is 2.3 $\pi$ mm-mrad. However, the PARMELA simulation gives an integrated rms emittance of 5.3 $\pi$ mm-mrad.

The discrepancy in emittance is due to the manner in which the data is analyzed. Measuring the full distribution of an image on a screen is susceptible to many errors. In particular, the correction of data due to baseline shifts and the non-linear response of cameras, especially at low intensity, is very difficult. Unfortunately, the rms emittance numbers are very sensitive to the tails of the distribution. So instead, many researchers measure an unambiguous parameter of the spot-size, the full-width half-maximum. As can be seen in Figure 30, the agreement between the FWHM's from the experimental measurement and the FWHM's from PARMELA is very good. However, because of the longitudinal dynamics of different slices, the FWHM measurement cannot be used to directly compute the beam emittance. In Figure 30, the dashed curve shows the FWHM as calculated from the rms spot sizes from PARMELA. By using the FWHM of the spot calculated from the rms emittance, the quadrupole scan fit gives an emittance close to the calculated emittance.

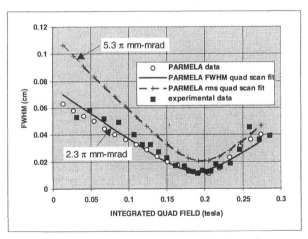

Figure 30. The data and PARMELA simulation are for a quadrupole scan with the FWHM taken at a screen 30 cm downstream from the quadrupole. The electron pulse is 1.9 nC at 17.2 MeV. The beam is produced by a $Cs_2Te$ cathode illuminated by a 8 ps laser pulse. The laser spatial profile is a 6 mm FWHM Gaussian clipped with a circle of 5.2 mm diameter. The FWHM of each slice at the screen is plotted as a function of the quadrupole gradient.

The variation of FWHMs at the screen of the individual slices with changing quadrupole strength is shown in Figure 31. The reason for the discrepancy in emittance is readily apparent. The ends of the micropulse are focused differently than the middle of the pulse. The FWHM spot size measurement is thus complicated by the different longitudinal portions of the pulse contributing to the FWHM in differing amounts as the quadrupole is varied.

The minimum spot size is dependent on the cathode temperature and any residual magnetic field on the cathode. Thus far, the cathode temperature of $Cs_2Te$ has not been measured. The cathode initial emittance can be inferred by adding a minimum spot size to the PARMELA spot sizes (square root of sum of squares). From the experimental data, this gives an initial emittance of 2.8 $\pi$ mm-mrad, corresponding to a transverse energy of 1.2 eV. The partition of this energy between residual magnetic field at the surface of the cathode and cathode temperature could not be determined.

Finally, the large solenoid around the cathode region is the main steering and focusing element in the system. This results in the beam's Twiss parameters, as well as the beam's propagation direction, being very sensitive to the magnitude and tilt of the solenoid's field. For the case shown above, the measured value of the large solenoid's field was within 1% (experimental error was +/-2%) of the value predicted by PARMELA.

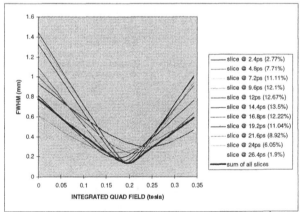

Figure 31. The TAPE2 PARMELA output was processed by dividing the longitudinal length of the pulse into 11 equal segments. The FWHM of each slice at the screen is then plotted as a function of the quadrupole gradient. The fraction of charge in each slice is shown in the legend. The thick black line is the summation of all the individual FWHM's. To make sure the slicing was done properly, the summation is compared with the normal output of PARMELA (an integration over all slices).

Changes in magnetic field as small as 1% are easily observable in simulation and have a significant effect on the beam's Twiss parameters.

## 6. APPLICATIONS

The two classes of applications emerge as the most likely for FELs are research and commercial applications. The application of FELs for research depends on whether an existing facility is used or a new facility is being built. If the FEL is added to an existing accelerator facility then the operations staff can be a sub-group of the existing operations staff. This significantly reduces costs since experts in subsystems are already present (such as controls and rf systems). Also, an FEL added to an existing facility is typically a small fraction of the overall facility cost, the FEL can use leftover and modified existing components, and the utilities are largely covered.

If a new FEL is being built for research activities, then the design of the FEL must be low-cost and have a design that does not require a large crew of experts. However, sophisticated operators are in good supply.

Finally, commercial applications have the most restrictive set of requirements. The FEL must have low initial-cost, have low-maintenance, and have low-operating cost. The FEL must also be user-friendly, i.e., employ one non-specialist operator. These requirements imply a high level (cost) of advanced engineering.

The next sections are a very brief summary of the National Academy of Science report on the applications of FELs. For more detailed information please refer to the National Research Council's report.[49] The report summary is divided into four sections based on wavelength: 1000 to 10 microns, 10 microns to 200 nm, 200 nm to 10 nm and x-ray wavelengths.

### 6.1 1000 to 10 microns

The 1000 to 10 micron wavelength band offers many opportunities for an intense, tunable light source for studies in surface science, chemistry, solid state physics, biophysics, and plasma physics. The following list indicates some potential applications not address by presently available sources.

Some surface science applications are the measurement of energy distributions and line shapes of intramoleculer vibrations, of the chemisorption of surfaces, and of the adsorption of species during chemical reactions. A chemistry application is the study of energy transfer in molecules in the gas and liquid phase. Some solid-state physics applications are: the measurement of phonon, plasmon, magnon, and inter-sub-band transition excitations in condensed matter, the direct probing of defect modes and buried interfaces, the probing of mode-mode interactions, and the driving of pre-selected strongly non-equilibrium states and studying their relaxation. A biophysics application is measurement of low-frequency modes of large biomolecules such as nucleic acids and proteins. Some plasma physics applications are plasma heating with tens of megawatts, plasma diagnostics, measuring phase shift of waves sent through a plasma, and measuring reflections of waves from critical density regions. A final application is isotope separation.

## 6.2  10 microns to 200 nm

In the 200 nm to 10 micron band, conventional lasers give stiff competition to free-electron lasers. The biggest problem is the cost comparison between FELs and conventional laser systems, with conventional systems costing between $5K to $250K.

An FEL's real advantage in this wavelength band is high-power at wavelengths not accessible to high-power conventional lasers. For instance, the batteries in satellites determine the satellite's operating lifetime. Power-beaming at 0.85 microns, the peak of the photocell response, would increase the satellite's lifetime several years. Another application for an FEL is in atmospheric science. Atmospheric modeling requires knowledge about the concentration of water and wind patterns between 20 km and 100 km. The conventional technique is to launch balloons. A high-power tunable laser could be used to sweep over waters rotational lines and get temperature information as well as concentration and wind speed as a function of altitude. Finally, a reliable, robust high-power laser would have defensive military applications.

## 6.3  200 nm to 10 nm

In the 10 nm to 200 nm band other sources are available by using 4-wave mixing of lasers or synchrotrons. Some possible FEL applications are: to study photodissociation dynamics by pumping molecules with the FEL and studying fragments with a tunable probe laser; to do photoelectron spectroscopy by probing many molecules (like radical and weakly bound complexes) not accessible now; to do pump-probe photoemission to study long-lived excited states in semiconductors for information on intrinsic and defect states; and for the commercial processing of polymers.

## 6.4 X-ray Wavelengths

In the 0.1 to 10 nm wavelength, the light sources do not have the peak power or coherence of an FEL. Competing light sources are synchrotrons, laser targets, Compton back-scattering. One application using time-correlation spectroscopy is measuring the time dependence of speckle patterns. This type of spectroscopy gives information about motion on less than 100 nm scales. A set of applications exists in the X-ray spectroscopy, such as: magnetic scattering at absorption edges to study magnetically ordered systems, inelastic x-ray scattering to probe high-energy phonons and magnons, and dynamical behavior of quasi-crystalline and fluid-phase short-range order. Another application is the microscopy and holography of biological process in cells.

Although not an FEL based system, Compton Back-Scattering (CBS) can be used as an efficient source tunable x-rays from a bright electron beam accelerator. The properties of a CBS system are angular dependent bandwidth and low electron beam energy (less than 25 MeV). If the accelerator is photoinjector-based, the time correlation of scattering laser and electrons is automatic. If the accelerator is part of a compact, low-cost system then two medical applications look promising: mammography and coronary angiography. These medical applications are interesting because a narrow x-ray linewidth tuned to the 33 keV edge of iodine would give better definition and at a greatly reduced dose.

The following is an example CBS system based on using the AFEL accelerator and assuming an electron beam energy of 22 MeV and a quadrupled Nd:YLF (4.6 eV) laser.[50] The back-scattering of the 4.6 eV photons off the electron beam gives x-rays at the 33-KeV edge of iodine. The x-ray photon energy, $E_{x\text{-ray}}$, is calculated by

$$E_{x\text{-ray}} = 4 * E_{laser} * E^2_{electron} / (mc^2)^2, \qquad (6.1)$$

where $E_{laser}$ is the photon energy, $E_{electron}$ is the electron energy, m is the electron mass, and c is the speed of light. The x-ray flux is

$$flux = N_e * N_{laser} * \sigma / A, \qquad (6.2)$$

where $N_e$ is the number of electrons, $N_{laser}$ is the number of laser photons, $\sigma$ is the cross-section, and A is the interaction area. Assuming a cross-section of 6.65 x $10^{-24}$ cm$^2$ and using a 5% bandwidth that corresponds to +/- 5 mrad, then each 20-ps electron micropulse gives 1000 x-ray pulses.

Each 30 microsecond macropulse has 3000 pulses, and the maximum repetition rate is 60 Hz giving a total of $1.8 \times 10^8$ photons/second.

## 6.5 Commercial Applications

There are many commercial applications of electron beams, however, almost all do not require high brightness electron beams. A sample list is: sterilization of medical products, food containers and products, sewage; decomposition/precipitation of chemical pollutants and hazardous materials; cancer therapy; cross-linking of polymers in insulation for wires and cables and polymerization; curing of composites, surface coatings, magnetic recording media, and adhesives; radiography; materials modification for ion implantation, gemstones, and radiation hardening.

## 6.6 High-Energy Physics

For high-energy physics the applications are reducing the acceptance requirements of a damping ring, and thereby reducing the cost, or possibly eliminating a damping ring completely. Of course, a ring would still be needed for the positrons. Another interesting application requiring both a high-brightness electron beam and an FEL is a gamma-gamma collider.[51]

## 7. SUMMARY

Photoinjector technology has had significant developments in the decade since its inception. Designs now span a large range in accelerator frequencies and electron pulse requirements. The photocathode source, though difficult, is not a major impediment to implementing a photoinjector-based system. However, the amplitude stability of the drive laser for the photocathode is an issue.

Design of a 20-MeV compact linac based on the photoinjector has been completed. The linac is approximately 1.2 m long and is operated with a 15-μs macropulse at up to 15 Hz with a 0.5-A average during the macropulse. The design of the linac is based on emittance reduction by reversing the effects of space charge after the photoinjector gun. An exact comparison with simulation is required for a thorough understanding of the phase space of the pulse. For a good simulation, an accurate measurement of magnetic fields, photocathode laser profile, accelerating fields, and phasing of the laser and rf is required. With accurate measurements, good agreement between the experiment and PARMELA simulations can be obtained.

The production of high-current high-brightness electron beams has enjoyed considerable progress over the last several years, mainly because of changes in the requirements imposed by free electron lasers. Several approaches show considerable potential for producing very bright electron beams. The concept of placing a photoemissive source in an accelerating structure has been demonstrated. The basic physics of photoinjectors is understood. Several groups around the world are designing bright beams based on this technology and continued improvement in photoinjector design is expected.

## ACKNOWLEDGMENTS

The author thanks John Fraser with whom the initial work on the photoinjector program was accomplished. Also, I thank Dominic Chan, Dinh Nguyen, John Kinross-Wright, Karl Meier, Lloyd Young, Rick Wood, Ed Gray, Tai-Sen Wang, Robert Springer, Bruce Lambertine, Valerie Loebs, Joel Johnson, Peter Oettinger, Bill Herrmannsfeldt, Roger Miller, Harold Hanerfeld, Bill Wilson, Bob Austin, Carl Timmer, Boyd Sherwood, Louis Rivera, John Plato, Mike Weber, Floyd Sigler, Steve Gierman, Steve Kong, Steve Russell, Steve Hartmann, Alessandra Lombardi, Dodge Warren, Cliff Fortgang, Bob Kraus, Doug Gilpatrick, and John Ledford for their contributions to the photoinjector projects. The author acknowledges Bruce Carlsten, Bill Herrmannsfeldt, Roger Miller, Charles Sinclair, Todd Smith, Steve Benson, Ken Batchelor, S. Chattopadhyay, and  R. Dei-Cas for their helpful discussions and for information on bright electron sources. The author is indebted to many individuals for information on their projects. In particular, I wish to thank Chris Travier, Steve Kong, John Adamski, Shien-Chi Chen, Harold Kirk, Claudio Pellegrini, and Jim Simpson. Also Jerry Watson and Stanley Schriber are acknowledged for their continued support.

## REFERENCES

1.  C. LeJeune and F. Aubert, *Applied Charged Particle Optics*, A. Septier, Ed., Advances in Electronics and Electron Physics, Supp. 13A, 159-259 (1980)
2.  K. L. Brown, *Nucl. Inst. And Meth.* **187**, 51-65 (1981)
3.  J. D. Lawson, *The Physics of Charged Particle Beams*, (Oxford Univ. Press, 1978) 175
4.  J. D. Lawson, *ibid.*, 199
5.  W. D. Herrmannsfeldt, *Physics of Particle Accelerators*, AIP Conf Proc. 184, **2**, 1533-1542 (1988)
6.  J. R. Pierce, *Theory and Design of Electron Beams*, (van Nostrand, 1949)

7.  T. I. Smith, *1986 Linear Conf. Proc.*, SLAC-303, 421-425 (1986)
8.  T. Energa, L. Durieu, D. Michelson, and R. Worsham, *IEEE Trans. Nucl. Sci.* **32** (5), 2936 (1985)
9.  S. V. Benson, J. Schultz, B. A. Hooper, R. Crane, and J. M. J. Madey, *Nucl. Instr. and Meth.*, **A272**, 22-28 (1988)
10. P. E. Oettinger and I. Bursuc, *IEEE Part. Accel. Conf,* IEEE cat. no. 87CH2387-9, **1**, 286 (1987)
11. J. Fischer, T. Srinivasan-Rao, and T. Tsang, *Sources '94*, Schwerin, Germany, Sept. 29 - Oct.4, 287-289 (1993)
12. G. Suberlucq, *Sources '94*, Schwerin, Germany, Sept. 29 - Oct.4, 557-561 (1993)
13. C. Travier, *6th Workshop on Advanced Accel. Concepts, Lake Geneva, WI, June 12-18,* (1994)
14. Private communication from M. E. Conde, UCLA/ANL (1994)
15. Private communication from C. Pellegrini, UCLA (1994)
16. D. J. Bamford, M. H. Bakshi, and K. A. G. Deacon, *Nucl. Inst. And Methods*, **A318**, 377-380 (1992)
17. Y. Tur, *Sources '94*, Schwerin, Germany, Sept. 29 - Oct.4, 572-576 (1993)
18. E.Chevallay, J. Durand, S. Hutchins, G. Suberlucq and M. Wurgel, *Nucl. Instr. and Meth.* **A340**, 146 (1994)
19. S. H. Kong, J. Kinross-Wright, D. C. Nguyen, and R. L Sheffield, *J. Appl. Phys.*, **77**, 1-8 (1995)
20. M. Yoshioka, *Sources '94*, Schwerin, Germany, Sept. 29 - Oct.4, 624-629 (1993)
21. J. S. Fraser, R. L. Sheffield, and E. R. Gray, *Laser Acceleration of Particles AIP Conf. Proc.*, no. 130, 598, (1985); J. S. Fraser, R. L. Sheffield, and E. R. Gray, *Nucl. Inst. And Methods*, **250**, 71-76 (1986)
22. B. E. Carlsten, *Nucl. Inst. And Methods*, **A285**, 313-319 (1988)
23. J. S. Fraser and R. L. Sheffield, *IEEE J. Quant. Elec.*, **QE-23**, 1489-1496 (1987)
24. W. Herrmannsfeldt, R. Miller, and H. Hanerfeld, SLAC-PUB-4663, (1988)
25. E. R. Gray and J. S. Fraser, *Proc. 1988 Linear Accel. Conf.*, Williamsburg, VA, October 3-7, CEBAF Report 89-001, 338-340 (1988)
26. S. V. Benson, J. M. J. Madey, E. B. Szarmes, A. Bhowmik, J. Brown, P. Metty and M. S. Curtin, *Nucl. Instr. and Methods*, **A296,** 762 (1990)
27. J. Early, J. Barton, G. Busch, R. Wenzel, and D. Remelius, *Nucl. Instr. and Meth.*, **A318**, 381-388 (1992)
28. R. L. Sheffield, R. H. Austin, K. D. C. Chan, S. M. Gierman, J. M. Kinross-Wright, S. H. Kong, D. C. Nguyen, S. J. Russell, and C. A. Timmer, *Proc. 1993 Part. Accel. Conf.*, **4**, 2970-2972 (1993)
29. D. H. Dowell, K. J. Davis, K. D. Fridell, E. L. Tyson, C. A. Lancaster, L. Milliman, R. E. Rodenburg, T. Aas, M. Bemes, S. Z. Bethel, P. E. Johnson, K. Murphy, C. Whelen, G. E. Busch, and K. K. Remelius, *Appl. Phys. Lett.*, **63**, no. 15, 2035-2037 (1993)
30. S. Joly et al., *Proc. 1990 European Part. Accel. Conf.*, 140-142 (1990)
31. K. Batchelor, I. Ben-Zvi, R. C. Fernow, J. Fischer, A. S. Fisher, J. Gallardo, G. Ingold, H. G. Kirk, K. P. Leung, R. Malone, I. Pogorelsky, T. Srinivasan-Rao, J. Rogers, T. Tsang, J. Sheehan, S. Ulc, M. Wookle, J. Xie, R. S. Zhang, L. Y. Lin, K.

T. McDonald, D. P. Russell, C. M. Hung, and X. J. Wang, *Nucl. Inst. and Methods*, **318**, 372-376 (1992)

32.  S. C. Chen, J. Gonichon, L. C-L. Lin, R. J. Temkin, S. Trotz, B. G. Danly, and J. S. Wurtele, *Proc. 1993 Par. Accel. Conf.*, **4**, 2575-2577 (1993)

33.  P. Schoessow, E. Chojnacki, G. Cox, W. Gai, C. Ho, R. Konecny, J. Power, M. Rosing, J Simpson, N. Barov, and M. Conde, IEEE Proc. of Part. Accel. Conf., May 1-5, Dallas, Tx, (1995)

34.  H. Kirk, *Sources '94*, Schwerin, Germany, Sept. 29 - Oct.4, 392-398 (1993)

35.  B. E. Carlsten, Porc.10th Int. FEL Conf., Jerusalem, Israel, 1988, *Nucl. Instr. and Meth.*, **A285**, 313 (1989); B. E. Carlsten and R. L. Sheffield, *Proc. 1988 Linac Conf.*, Williamsburg, Va, 1988, CEBAF Report 89-001, 365-369 (1989); B. E. Carlsten, *Proc. 1989 IEEE Part. Accel. Conf.*, Chicago, IL, IEEE Catalog no. 89CH2669-0 313 (1989); B. E. Carlsten, *Part. Accel.*, **49**, 27-65 (1995)

36.  B. E. Carlsten, *Part. Accel.*, **49**, 34-37 (1995)

37.  K. C. D. Chan, R. H. Kraus, J. Ledford, K. L. Meier, R. E. Meyer, D. Nguyen, R. L. Sheffield, F. L. Sigler, L. M. Young, T. S. Wang, W. L. Wilson, and R. L. Wood, *Nucl. Instr. and Methods*, **A318**, 148-152 (1992)

38.  P. G. O'Shea, S. C. Bender, S. A. Byrd, B. E. Carlsten, J. W. Early, D. W. Feldman, R. B. Feldman, W. J. D. Johnson, A. H. Lumpkin, M. J. Schmitt, R. W. Springer, W. E. Stein and T. J. Zaugg, *Nucl. Instr. and Methods*, **A318**, 52-57 (1992)

39.  T. F. Wang, K. C. D. Chan, R. L. Sheffield, and W. L. Wilson *Nucl. Instr. and Methods*, **A318**, 314-318 (1992)

40.  W. B. Colson, G. Dattoli, and F. Ciocci, *Phys Rev. A*, **64**, 2 (1985)

41.  S. Hartmann, A. Lombardi, R. L. Sheffield, and T. Wang, *Nucl. Instr. and Methods*, **A318**, 319-322 (1992)

42.  D. C. Nguyen, R. H. Austin, K. C. D. Chan, C. Fortgang, W. J. D. Johnson, J. Goldstein, S. M. Gierman, J. Kinross-Wright, S. H. Kong, K. L. Meier, J. G. Plato, S. J. Russell, R. L. Sheffield, B. A. Sherwood, C. A. Timmer, R. W. Warren, and M. E. Weber, *Nucl. Instr. and Meth.*, **A341**, 29-33 (1994)

43.  K. C. D. Chan, K. L. Meier, D. C. Nguyen, R. L. Sheffield, T. S. Wang, R. W. Warren, W. Wilson and L. M. Young, *1992 Linear Accelerator Conf. Proc.*, Ottawa, Ontario, Canada, AECL-10728, **1**, 37-39 (1992)

44.  J. D. Gilpatrick et al., *IEEE Proc. of Particle Accelerator Conf.*, (1993)

45.  D. W. Preston and R. W. Warren, *Nucl. Instr. and Methods*, **A318**, 794-797 (1992)

46.  M. Xie and K. J. Kim, *Nucl. Instr. and Meth.*, **A318**, 877-884 (1992)

47.  B. E. Carlsten, J. C. Goldstein, P. G. O'Shea, and E. J. Pitcher, *Nucl. Inst. and Methods*, **A331**, 791-796 (1993)

48.  S. Humphries, Jr., "Charged Particle Beams," *John Wiley and Sons, Inc.*, 143 (1990)

49.  National Reasearch Council, "Free-Electron Lasers and Other Advanced Sources of Light," Board on Chemical Sciences and Technolgoy, National Reasearch Council, 2101 Constitution Ave. N.W., Washington, D.C., 20418 (1994)

50.  D. C. Nguyen, S. M. Gierman, W. Vernon, and R. L. Sheffield, *Nucl. Inst. and Methods*, **A358**, ABS 48- ABS 49 (1995)

51.  A. M. Sessler, "Photon-Photon Colliders," *IEEE Proc. of Particle Accelerator Conf.*, Dallas, TX, May 1-5 (1995)

# APPLICATIONS OF FREE-ELECTRON LASERS

NORMAN H. TOLK, R. G. ALBRIDGE, A. V. BARNES, B. M. BARNES, J. L. DAVIDSON, V. D. GORDON, G. MARGARITONDO*, J. T. McKINLEY, G. A. MENSING, J. STURMANN
Center for Molecular and Atomic Studies at Surfaces, Department of Physics and Astronomy, Vanderbilt University, Nashville TN 37235 USA
*Ecole Polytechnique Fédérale, CH-1015 Lausanne, Switzerland

**Abstract**    Free-electron lasers (FEL's) are important tools for producing high-intensity photon beams, especially in the infrared. Synchrotron radiation's primary spectral domains are in the ultraviolet and x-ray region. FEL's are therefore excellent complimentary facilities to synchrotron radiation sources. While FEL's have seen only limited use in experimentation, recently developed programs at Vanderbilt University in Nashville, the University of California at Santa Barbara, the LURE laboratory in Orsay, the FOM in Holland, and elsewhere are swiftly rectifying this situation. This paper examines practical experience obtained through pioneering programs using the Vanderbilt FEL, which currently hosts one of the largest materials research programs with a FEL as well as significant medical research programs. Results will be discussed in four areas: two-photon absorption in germanium, FEL-assisted internal photoemision measurements of interface energy barriers (FELIPE), wavelength-specific laser diamond ablation, and tissue ablation based on the FEL-produced resonant denaturation of structural proteins.

## 1. INTRODUCTION

The free-electron laser (FEL) is not only a technological achievement in laser design, but it also is a powerful instrument which offers researchers the capacity to conduct research using high-intensity photon beams. As synchrotron radiation sources have been used for many experiments in the ultraviolet and x-ray regions, the FEL is an excellent source for a wide array of infrared-photon projects and applications. This paper is not intended to be a comprehensive listing of all FEL applications. Rather, the two broad uses presented here are merely examples of applications of the FEL at Vanderbilt University. One application employs the FEL in spectroscopy as a probe of electronic and vibrational structures, as has been performed in measurements of internal photoemision of interface energy barriers (FELIPE) and in the study of two-photon absorption in germanium. The FEL also provides unique opportunities to research the selective alterations of materials. Two examples of materials modification are wavelength-specific laser diamond ablation and tissue ablation based on the FEL-produced resonant denaturation of structural proteins.

## 2. INTERNAL PHOTOEMISSION

The infrared radiation emitted by the Vanderbilt FEL offers a chance to resolve important problems in interface physics. For many years, concentrated experimental and theoretical research has focused on the band lineup at the interface of two semiconductors [1]. The accuracy in measuring interface energy barriers can be improved to the level required to test current theories. Such tests can be performed on buried interfaces rather than on those formed by thin overlayers. Internal photoemission (IPE), an optical technique with very high accuracy, is a simple and direct method of determining the band discontinuities at semiconductor interfaces. Other techniques require complex modeling in order to interpret the experimental data, but IPE does not. We have recently shown the feasibility of conducting semiconductor heterojunction band discontinuity measurements of ultrahigh accuracy using the free-electron laser internal photoemission (FELIPE) on AlGaAs/GaAs[2] and GaAs/amorphous-Ge heterojunctions [3].

Using the FELIPE technique, photons are used to pump electrons across the energy barrier caused by a heterojunction band discontinuity. We derive the barrier height directly from the spectral position of the threshold. An alternate method uses a conventional photon source. The threshold corresponds to the sum of the discontinuity plus the local gap width. This method can provide information on the local optical properties of the materials in the junction and can also check the results of the direct FEL measurements. This enhances the reliability of the FEL measurements and thus reduces a serious problem in other interface barrier measurements.

The Vanderbilt University FEL is an upgraded version of the Stanford University Mark III FEL[4]. Tunable over the 2-10 $\mu$m wavelength, it can reach 1 $\mu$m using frequency doubling. The FEL can produce sequences of 5-10 MW pulses with duration of ~700 fs at a repetition rate of 2.9 GHz. A typical series has ~5 $\mu$s duration, 200 mJ pulse energy, 30 Hz maximum repetition rate, ~5% pulse-to-pulse energy fluctuations, and 0.7% fundamental linewidth.

The Ga/As/a-Ge heterojunctions used in these experiments were prepared[5] by evaporating amorphous-Ge layers of ~5 mm diameter and 1 $\mu$m thickness onto an n-type, single-crystal GaAs (110) substrate. An Au dot of ~2 mm diameter was deposited near the perimeter of the Ge layer in order to achieve ohmic contact. A grid of Ni/Ge/Au patterns was back-contacted to the GaAs substrate. The substrate was then etched using standard fabrication processes. At 77 K, the forbidden gap values are 1.51 and 0.75 eV for the GaAs and a-Ge. The measurements were performed in an optical cryostat at 77 K. The cryostat was configured so that the FEL beam entered the heterojunction through the a-Ge layer. The undesired higher harmonics of the FEL were filtered out. Using a pyroelectric joulemeter, the photocurrent was normalized to the laser power. While one side of the heterojunction was held at the bias voltage, the other was connected to a Keithley 428 fast current amplifier.

The FEL produces an observable photocurrent by photo-assisting electrons over the conduction band discontinuity as shown in Figure 1a. A representative IPE spectrum is given in Figure 1b, showing how $\Delta E_c$ is given directly by the photon energy at the photocurrent threshold. A variety of theoretical lineshapes have been suggested for the near-threshold region[6], all of the form $Y \propto (h\nu - \Delta E_c)^n$, where Y is the photocurrent yield, hν is the photon energy, and n is 2, 2.5, or 3 in the Fowler, Kane, and Powell models. When the Fowler fit is replaced by a simple linear fit (n=1), the estimate of $\Delta E_c$ changes by only 5 meV. For biases of the a-Ge side relative to the GaAs side of -0.8 V and -1 V, the linear fit gives $\Delta E_c$ = 0.399 eV while the Fowler fit gives 0.334 eV. For bias voltages between 0 and -1 V, no significant variations of $\Delta E_c$ were found. This indicates that tunneling corrections are negligible. A 100 V bias was applied across two Ni/Ge/Au back contacts on the GaAs. No threshold was observed in the resulting photoconductivity spectrum; therefore the threshold in the IPE spectrum is not a bulk GaAs effect.

IPE can be used to study interfaces buried deep beneath a surface, unlike conventional photoemission spectroscopies [1] which require the interface to be within a few monolayers of the surface. This makes it possible to study device-quality interfaces without cumbersome in-situ growth being required. FELIPE is also complementary to earlier GaAs/a-Ge IPE experiments [5]; these experiments induced transitions from the a-Ge valence band maximum to the GaAs conduction band minimum using monochromatic visible light. In this case $\Delta E_c$ is given by the threshold photon energy less the a-Ge band gap. As long as flat-band conditions exist at the interface [7], a direct measurement of the interfacial band gap of the small gap material can be obtained by comparing the results of visible-IPE with those of FELIPE. Since processes such as microdiffusion and interfacial strain introduce uncertainty about the value of the near-interface band gap relative to the bulk gap, this is important information. The near-interface and bulk gaps agreed to within experimental error for the case of GaAs/a-Ge.

The first experimental data obtained with this method enabled scientists from the Centre de Spectromicroscopie to measure the discontinuity of a GaAl/AlGaAs heterojunction with high accuracy. Recent data has been of even higher quality. At this level of accuracy, one can

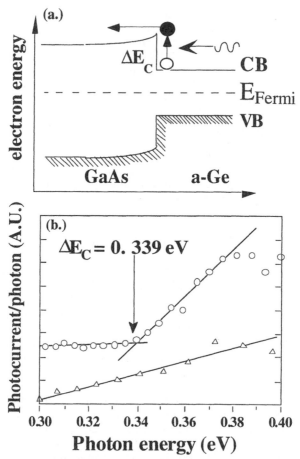

FIGURE 1   (a.) Energy band diagram of the GaAs/a-Ge interface. If the photon energy exceeds $\Delta E_c$, then a photocurrent is produced. (b.) Upper lines - linear fit to threshold region for GaAs/a-Ge interface with -0.8 V bias. lower line - photocurrent measured by applying 100 V between two electrodes attached to the Ni/Ga/Au back contacts of the GaAs, showing that the photocurrent threshold is not due to bulk GaAs photocurrent.

easily, directly, and reliably assess the barrier height with an accuracy which varies somewhat between experiments depending on specific experimental conditions but always lies in the range of a very few meV.

## 3. TWO-PHOTON ABSORPTION

The development of the free-electron laser allows the study of difficult spectroscopic problems unsolvable by conventional laser light. Two-photon spectroscopy has long been an important component of solid-state research due to its selection-rule complementarity with respect to conventional optical spectroscopy and also because two-photon absorption demonstrates many interesting and nontrivial features. While certain aspects of this spectroscopy could be practically implemented after the invention of lasers [8], two-photon absorption in germanium had never been thoroughly studied. The last measurements of this sort were performed in 1976 using conventional lasers; these did not reach the direct gap and no measurement was available of two-photon absorption at the indirect gap [9]. The FEL provides the technology needed to achieve a positive test of the decades-old Bassani-Hassan predictions.

The results of our Ge two-photon absorption experiments [10] show that the coefficient of indirect two-photon absorption is three orders of magnitude less than the direct gap coefficient. They also demonstrate that indirect absorption is LO-phonon assisted, which provides a positive test of the Bassani-Hassan predictions [11-13].

These results are based on photoconductivity measurements at ~125 K. Photoluminescence measurements conducted at ~7 K are discussed elsewhere. The sample used for photoconductivity measurements was a $p^+$-i-$n^+$ diode in an EO-817P North Coast Co. photodetector. The intrinsic region, 4 mm thick, was made of high-purity, low-defect-density Ge, with a net carrier concentration of $10^9$-$10^{10}$ cm$^{-3}$. The $p^+$-i-$n^+$ structure was biased at -300m V in order to fully deplete the intrinsic region before the photoconductivity shown in Figure 1 was measured. Photoconductivity was converted into a voltage change, $\Delta V$, by a transimpedance amplifier. The voltage change was measured by a boxcar integrator. An Au-doped Ge detector provided a reference FEL intensity signal $V_r$. Au impurity levels within the Ge gap give the detector broad spectral sensitivity dominated by single photon absorptions. This reference signal eliminated the effects of FEL intensity fluctuations. Composite data from three different runs was plotted as $\Delta V/V_r^2$ vs. twice the photon energy, $2\hbar\omega$. Fits of the data show that, as expected for two-photon processes, the near-threshold intensity dependence is essentially quadratic. The data in Fig. 2 gives a direct gap of $E_g^d = 0.87 \pm 0.01$ eV. Using Mahan's excitonic theory [14], with $E_g^d$ as an adjustable parameter, 0.87 was again obtained, confirming the previous evaluation. A gap of 0.87 eV corresponds to approximately 125 K, as this value is consistent with the published value [15,16] corrected for the sample temperature.

The value of $E_g^d$ obtained from this data analysis was then used to analyze the indirect gap data. It can be safely assumed that the difference between the gaps is independent of temperature, as the published dependence [15,16] is negligible within the experiment's level of accuracy. From the indirect gap width [15,16] at 77K of 0.737 eV and the direct-gap decrease of 19 meV from 77-125 K, an indirect gap width $E_g^i = 0.718 \pm 0.01$ eV at 125 K was derived.

Figure 2b clearly shows that the two-photon indirect-gap threshold is clearly shifted with respect to $E_g^i$. This shift is consistent with a process assisted by the emission of a 30.4 meV LO (L6-) phonon [16]. As Bassani and Hassan [11] predicted in 1972, therefore, the two-photon indirect

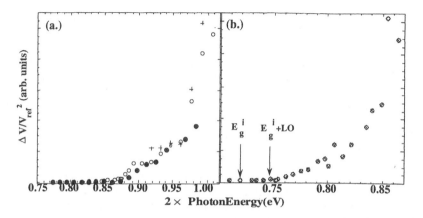

FIGURE 2    Two-photon absorption induced photoconductivity signal as a function of double the incident photon energy, with normalization to the reference signal $V_{ref}$, (a.) Threshold indicating a direct gap width of 0.87 eV at 125 K. (b.) Similar data for the indirect gap. The pointers mark the predictions of the Bassani-Hassan theory for the indirect gap width $E_{gi}$, and the corresponding LO phonon-assisted process.

threshold is primarily an LO-assisted process, differing from an one-photon mechanism which can be assisted by several different types of phonons. Several authors [17-19] have measured the absolute value of the two-photon absorption coefficient $\beta$ in the 0.34-2.5 cm/MW range above the direct gap, where $\beta$ is defined [8, 17, 20] by the relation $(dI/dz) = -\beta I^2$ and $I(z)$ is the intensity as a function of sample depth. Using our data to estimate the relative strength of the direct and indirect absorptions, we found that the indirect absorption was weaker by a factor of ~2 * $10^3$. We reached this approximation by determining the ratio of the absorption at a photon energy of 0.490 eV (above the two-photon direct-gap threshold) to the absorption at a photon energy of 0.407 eV (between the indirect and the direct gaps) and assuming that only two-photon absorption contributes to the photocurrent. This assumption is consistent with the earlier evidence by Gibson et al. [17] that one-photon contributions are negligible for the beam intensities used in our experiments and for a crystal of this size and impurity concentration.

## 4.  LASER ABLATION OF DIAMOND

The Vanderbilt FEL is also a useful device when used for the selective alterations of materials. In recent years, increasing interest has accrued to the clarification of photon-induced resonant bond-breaking mechanisms because this method can achieve materials alteration while subjecting the sample to only a fraction of the energy required in a thermal bond-breaking process. We have performed wavelength-dependent laser ablation studies on 10 μm-thick chemical vapor deposition (CVD) diamond films [21] using the FEL. Near the 3.5 μm wavelength, a sharp dip was observed in the ablation threshold. This correlates strongly to the 3.47 μm C-H stretch absorption band.

Measurements were conducted in a high vacuum of $10^{-9}$ Torr. The profile of the FEL is Gaussian; after focusing, the focal spot region, with intensity greater than $e^{-2}$ of the beam center intensity, is $\approx 56$ μm in diameter for a wavelength of 3.5 μm. Holes with diameters of 45 μm were produced by pulse energies far above the damage threshold; this agrees roughly with the above definition of the focal spot diameter. A single FEL pulse was directed at the sample and the diameter of the resulting hole was measured using a scanning electron microscope (SEM), thus determining the ablation threshold. A sharp minimum of 23 GW/cm$^2$ at 3.5 μm was found for the ablation threshold. An infrared absorption spectrum was taken using an FTIR with p-polarized light was incident at Brewster's angle relative to the diamond film in order to eliminate interference between multiple reflections. Multiple reflections occur because the film thickness (10 μm) and infrared wavelength (~3.5 μm) have similar magnitudes.

Furthermore, several stages of laser-induced modification can be observed in the SEM micrographs as a function of intensity prior to formation of holes in the film. A small white spot in the SEM micrograph, indicating a diminished electrical conductivity, is the first sign of damage. Craters surrounded by small plasma-effected areas appear at higher pulse intensities. The craters become holes through the entire thickness of the film at still higher intensities. The white spots represent the pre-graphitization phase, which is of particular interest and may represent transformation of graphite or amorphous carbon into an sp$^3$ phase at the grain boundaries.

As well as SEM micrographs, we took spatially-resolved Raman spectra with lateral resolution of 5 μm at various positions relative to the center of an ablation hole. A sharp Raman line at 1332 cm$^{-1}$, which is characteristic of diamond, was observed in the undamaged part of the film far from the hole. Also, we detected a very weak, broad band between 1400 and 1600 cm$^{-1}$; this is characteristic of the amorphous carbon which exists at the grain boundaries of the polycrystalline diamond film. The broad feature characteristic of amorphous carbon totally dominates the sharp diamond Raman line in the region from 25 to 50 μm. Two peaks characteristic of graphite (the 1580 cm$^{-1}$ G-band and the 1350 cm$^{-1}$ D-band) dominate the Raman spectrum within a few microns of the hole's edge; no diamond feature remains. Clearly, complete graphitization must precede large-scale ablation.

The initial onset of desorption was measured using a much more sensitive technique based on time-of-flight (TOF) spectroscopy. Using this technique, the ablation threshold intensity was determined to be much larger than the initial photodesorption threshold. The TOF spectrometer is a 50 cm magnetically shielded drift tube which is terminated by a mesh and a Chevron microchannel plate detector. Between the mesh and the microchannel plate there is a 200 V accelerating potential. The beam was incident at 60° relative to the sample surface normal. The drift tube lay in

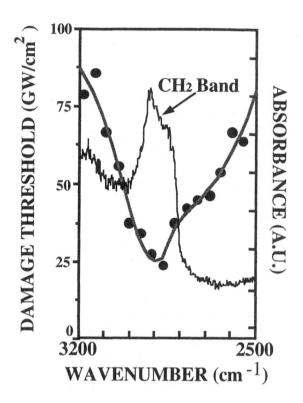

FIGURE 3  A comparison of the wavenumber dependence of the laser ablation threshold for CVD-growth diamond in the vicinity of the C-H stretch infrared absorption band and a FTIR spectrum of the C-H stretch absorption band for the same sample material (solid line with points), showing the minimum ablation threshold corresponds to the peak C-H stretch absorption.

the sample's plane of incidence and was perpendicular to the FEL beam. The drift time was measured with respect to the onset of the pyroelectric joulemeter signal. The particle mass was identified using a quadrupole mass analyzer.

A sharp increase in the desorption at an intensity of only 3 $GW/cm^2$ was observed. This is a much smaller intensity than the 23 $GW/cm^2$ found for the 3.5 μm ablation threshold. $C_2$ was found to be the dominant desorbed species. A large peak occurs in the TOF spectra at 130 μs drift time for diamond at low FEL intensities. This feature can be attributed to $C_2^+$, which has a corresponding particle energy of about 1.85 eV. For higher FEL intensities, we have observed a shift to shorter drift times, which correspond to higher particle energies. This post-ablation effect occurs because ion-ion repulsion increases in the increasingly energetic plasmas [21]. Quadrupole mass spectrometry demonstrates that the dominant desorbed ion is $C_2^+$. Furthermore, optical plasma fluorescence yielded that the dominant excited neutral was $C_2^*$. At 100 μs drift time, an extra TOF feature appears if higher FEL intensities are used. This unspecified feature may be compared to peaks in the graphite spectra at ~80 μs and ~110 μs. Another contrasting property discovered in this experiment was the desorption rate of the diamond and graphite; graphite films showed gradual increases starting at an intensity of 0.1 GW/cm², while diamond films showed a sudden increase at 3 GW/cm².

## 5. LASER ABLATION OF TISSUE

The Vanderbilt FEL is not only used for altering diamond films but also for altering tissue. Previous research in soft tissue ablation focused on the OH-stretch mode of water, which is in the infrared at about 3.0 μm [22-24]. However, experiments showed that collateral (thermal) damage and possible photochemical effects accompanied this approach to tissue ablation. Using the Vanderbilt FEL, Edwards et al. irradiated tissue with light tuned to the amide II band of proteins. This laser light, which has a wavelength of 6.45 μm and a peak power in the megawatt range, caused minimal collateral damage while providing an adequate ablation rate, as seen in Figure 4 [25].

Ocular and neural tissues were targeted to test the precision of the ablation process and dermis tissue was used to measure the efficacy of macroscopic incisions. Because the results of this research cannot be placed easily within the two popular models of tissue ablation (optical-breakdown and thermal-confinement), another model has been proposed. The partitioning-of-energy model specifies that collagen in the tissue may be denatured as structural transitions from highly ordered arrays to an

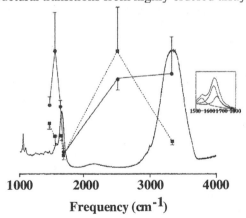

FIGURE 4   Relative ablation yield(●) and collateral damage(■)plotted against the absorption spectrum for corneal stroma. Reprinted with permission from Nature [25] Copyright 1994 Macmillan Magazines Limited.

58 N.H. TOLK *et al.*

amorphous gelatin occur. This process would take place at energies lower than the activation energy for depolymerization and at temperatures less than the temperature for pyrolytic fragmentation of biopolymers.

## 6. CONCLUSION

The Vanderbilt Free-Electron Laser's tunability (2-10μm), high intensity (15MW) and short pulse structure (1ps) make it ideal for studying (a) the electronic and vibrational structure of adsorbed surface molecules, and small and wide band-gap semiconductors, and (b) wavelength-selective materials alteration of both conventional and living state materials.

## ACKNOWLEDGMENTS

This work was supported by the Office of Naval Research under Grants N00014-91-J4040 and N00014-91-C-0109. We gratefully acknowledge the support of Charles Brau, Marcus Mendenhall, and the entire staff of the Vanderbilt Free-Electron Laser Center, without whom this research would not have been possible.

## REFERENCES

1. F. Capasso and G. Margaritondo, Heterojunction Band Discontinuities: Physics and Device Applications, North Holland, Amsterdam, 1987)
2. C. Coluzza, E. Tuncel, J.-L. Staheli, P. A. Baudat, G. Margaritondo, J. T. McKinley, A. Ueda, A. V. Barnes, R. G. Albridge, N. H. Tolk, D. Martin, F. Morier-Genoud, C. Dupuy, A. Rudra, and M. Ilegems, Phys. Rev. B 46 (1992) 12834
3. J. T. McKinley, R. G. Albridge, A. V. Barnes, A. Ueda, N. H. Tolk, D. Martin, F. Morier-Genoud, C. Dupuy, A. Rudra, and M. Ilegems,.J. Vac. Sci. Technol. B 11 (1993).
4. N. H. Tolk, C. A. Brau, G. S. Edwards, G. Margaritondo, and J. T. McKinley, Proc. Conf. on Short-Wavelength Radiation Sources, San Diego, California, 1991, SPIE Porc. Ser. Bellingham, Washington, USA, (1991) Vol. 1552, p. 7.
5. C. Coluzza, A. Neglia, A. Bennouna, M. Capizzi, R. Carluccio, A. Frova, and P. C. Srivastava, Appl. Surf. Sci. 56-58 (1992) 733.
C. Coluzza, F. Lama, A. Frova, P. Perfetti, C. Quaresima, and M. Capozi, J. Appl. Phys. 64 (1988) 3304;
6. R. H. Fowler, Phys. Rev. 38 (1931) 45;
E. O. Kane, Phys. Rev. 147 (1966) 335;
R. J. Powell, J. Appl. Phys. 41 (1970) 2424.
7. G. Abstreiter, U. Prechtel, G. Weimann, and W. Schlapp, Physica 134B (1985) 433;
G. Abstreiter, U. Prechtel, G. Weismann, and W. Schlapp, Surf. Sci. 174 (1986) 312.

8. For an extensive review of the experiments in this field, see: V. Nathan, A. H. Guenther, and S. S. Mitra, J. Opt. Soc. Am. B 2 (1985) 294.

9. A. F. Gibson, C. B. Hatch, P. N. D. Maggs, D. R. Tilley, and A. C. Walker, J. Phys. C 9 (1976) 3259.

10. E. Tuncel, J. L. Staheli, C. Coluzza, G. Margaritondo, J. T. McKinley, R. G. Albridge, A. V. Barnes, A. Ueda, X. Yang, and N. H. Tolk, Phys. Rev. Lett. 70 (1993) 4146.

11. F. Bassani and A. R. Hassan, Nuovo Cimento 7B (1972) 313.

12. F. Bassani and G. Pastori-Parravicini, Electronic States and Optical Transitions in Solids (Pergammon, Oxford 1975).

13. A. R. Hassan, Nuovo Cimento 13B (1973) 19.

14. G. D. Mahan, Phys. Rev 170 (1968) 825.

15. T. P. McLean, in: Progress in Semiconductors, ed. A. F. Gibson (Heywood, London, 1960) Vol. 5, p. 53.

16. Landolt-Börnstein, Numerical Data and Functional Relationships in Science and Technology, eds. M. Cardona, G. Harbeke, A. Rössler, and P. Madelung (Springer, Berlin 1982), Gp. III, Vol. 17a, p. 87.

17. A. F. Gibson, C. B. Hatch, P. N. D. Maggs, D. R. Tilley, and A. C. Walker, J. Phys. C 9 (1976) 3259.

18. B. V. Zubov, L. A. Kulevskii, V. P. Makarov, T. M. Murina, and A. M. Prokohorov, Zh. Eksp. Teor. Fiz. Pis'ma Red. 9 (1969) 221 [Sov. Phys. JETP Lett. 9 (1969) 130].

19. R. G. Wenzel, G. P. Arnold, and N. R. Greiner, Appl. Opt. 12 (1973) 2245.

20. C. C. Lee and H. Y. Fan, Phys. Rev. B 9 (1974) 3502.

21. A. Ueda, J. T McKinley, R. G. Albridge, A. V. Barnes, N. H. Tolk, J. L. Davidson, and M. L. Languell, Mater. Res. Soc. Symp. Proc. 285 (1993) 215; N. H. Tolk, R. G. Albridge, A. V. Barnes, J. T McKinley, A. Ueda, J. F. Smith, J. L. Davidson, M. L. Languell, C. Coluzza, E. Tuncel, and G. Margaritondo, Proc. SPIE 1854 (1993) 60.

22. R. Srinivasan, Science 234 (1986) 559-565.

23. B. Zysset, J. G. Fujimoto, C. A. Pulliafito, R. Birngruber, and T. F. Deutsch, Lasers Surg. Med. 9 (1989) 193-204.

24. J. T. Walsh Jr., T. J. Flotte, and T. F. Deutsch, 9 (1989) 314-326.

25. D. Stern, C. A. Pulliafito, E. T. Dobi, and W. T. Reidy, Ophthalmology 95(10) (1988) 1434-1441.

26. G. Edwards, R. Logan, M. Copeland, L. Reinisch, J. Davidson, B. Johnson, R. Maciunas, M. Mendenhall, R. Ossoff, J. Tribble, J. Werkhaven, and D. O'Day, Nature 371 (1994) 416-419.

# FREE ELECTRON LASER RESEARCH IN CHINA

JIALIN XIE
Institute of High Energy Physics, Chinese Academy of
Sciences, P.O.Box 918, Biejing, 100039, China

**Abstract**    China has endeavored in the field of FEL research since mid-eighties. Compton regime, Raman regime, Electromagnetic Pumping, Cherenkov FEL, and Harmonic generator FEL etc. based on   rf linac accelerator, induction linac accelerator, pulsed-line accelerator, and storage ring have all been developed. These experimental projects are in various stages of developments and will be discussed in sequence. Theoretical analysis, numerical simulation  and components research related to FEL will also be mentioned briefly.

## INTRODUCTION

Free Electron Laser is well known for its stringent requirements on the quality of the electron beam  taking part in the beam-wave interaction.  Thus research on FEL serves both the purposes of exploration of its potential applications and promotion of the developments of accelerator physics and technology.

FEL research and development were started in China since mid-eighties.[1] The primary purpose  is  to study  the physics and learn  the technology . Now, after some facilities had been completed and lased to saturation,  the applications   of these facilities become the object of exploration. Furthermore, in order to satisfy the requirements of various applications, the quality of the laser produced, such as the intensity and spectral stability etc. and the ease of operation of the system will come to attention and  deserve research efforts.

In the following, RF linac based Compton regime FEL, including Beijijng

IR-FEL of the Institute of High Energy physics; FIR-FEL of the Institute of Atomic Energy; wide-band user facility of the Institute of Nuclear Physics, and FIR-FEL of the Academy of Engineering Physics will be presented first. Then the Raman regime FEL amplifier of the Academy of Engineering Physics and E.M. pumped oscillator and Cherenkov FEL of several Institutes will be discussed. Finally, the UV harmonic generator of China University of Science and Technology will be described. Theoretical achievements, computer-code developments and system components studies will also be briefly mentioned.

## (1)  IR-FIR FEL

### (1.1)  IR-FEL of  The Institute of High Energy Physics (BFEL) [ 2 ]

Fig.1 shows the schematic layout of BFEL. This facility uses micrrowave electron gun as electron source with $LaB_6$ cathode of <100> cut for high stability and high emission. The maximum energy of the electrons from the gun cavity is 1.2 Mev and the beam current is about 200ma at the entrance

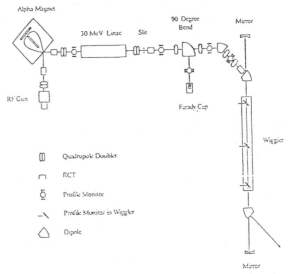

FIGURE 1   Schematic Diagram of Beijing FEL

of the linac with a bunch width of about 4 ps. The thermionic cathode wicrowave electron gun proves to be a very compact and convenient way for the production of short electron bunch with good beam quality for IR-FEL. However, it has the draw-back of back-bombardment that cause intensity variation of the beam current during the macropulse accompanied by energy variation due to the beam loading effect in the gun cavity. By using two deflecting magnetic fields along the gun cavity, the back-bombardment effect can be significantly reduced. BFEL operates stably at a macropulse length of $4.5\,\mu\,\mu$s and repetition rate of $3.125$Hz.

The electrons generated from the gun and compressed by the Alpha magnet  are injected into a S-band ,constant gradient ,unit relative phase velocity linac section and is accelerated to about 30 Mev. The linac section used is a modified Slac type constant gradient waveguide with four extra circular holes on some disks. These holes create a separation of the dispersion curves of $HEM_{11}$ mode but little purterbation of the dominant $TM_{01}$ mode.[3] Thus the BBU threshold value should increase with practically no effect on the acceleration process.

After passing through an achromatic and nearly isochronous 90° beam transport system, electron bunches are injected into an undulator.The undulator is planar type with NdFeB permanent magnets, consisting of 50 periods of 1.5m long. The magnets have the size of 7.5mm*7.5mm*40mm. The optical cavity is a near concentric resonator consisting of two ZnSe mirrors with ZnSe/TuF4 multilayer dielectric coating. After extracted from the downstream mirror, the laser beam passes through a telescopic pipe and is sent to the diagnostic room.

. The typical results of lasing is illustrated in Fig.2. where  the laser builds up in about 2µs, corresponding to about 100 round trips in the optical cavity The saturated output power  power level is about 2 Kw at the HgGdTe detector. since both the outcoupling factor and the macropulse duty factor are about 1%, the intra-cavity average optical power is estimated to be 200Kw and the peak power , 20 Mw. The spacial distribution of the laser at the diagnostic room, as measured with a LMP-32*36 elements pyroelectric detector located in the focal plane of a lens, is about the diffraction limit as illustrated by Fig.3. The wave-length range of the system was so far limited by the spectral range of the reflectance of the mirror. The system parameter of BFEL is in Table 1.

TABLE 1. BFEL system parameters

| | |
|---|---|
| **Electron beam** | |
| Macropulse length | 4.5 $\mu$s |
| Macropulse repetition rate | 3.125 Hz |
| Macropulse length | 3-4ps |
| Micropulse repetition rate | 2856MHz |
| Beam energy | 24-28Mev |
| Energy spread (FWHM) | 0.7% |
| Macropulse current | 150-200ma |
| Normalized emittance at rf gun exit | 20 $\pi$mm-mrad |
| **Undulator II** | |
| Period | 3cm |
| Number of periods | 50 |
| Gap | 1.15cm |
| K value | 1.17 |
| Electron trajectory deviation | 50 $\mu$m |
| Harmonic contents | 1% |
| **Optic cavity** | |
| Cavity length | 251.9cm |
| Operating wavelength | 9-11 $\mu$m |
| Mirror radii, up stream | 174cm |
| down stream | 170cm |
| Mirror reflectance, up stream | 99.5% |
| down stream | 99% |
| Rayleigh length | 76.5cm |

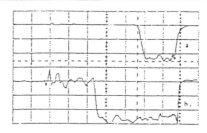

FIGURE 2. BFEL macropulse at Saturation. ( 1μs/ div ). (a) laser; (b) electron beam.

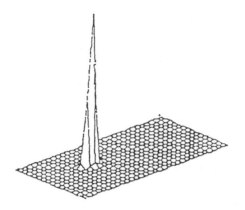

FIGURE 3. Spatial distribution of the laser beam

TABLE 2. Summary of the laser characteristics

| | |
|---|---|
| Total spectral range | 9-11 $\mu$m |
| Output energy in optical macropulse | 2-10mJ |
| Average power in macropulse | 200KW |
| Peak power | 20MW |
| Spectral width | 0.3-2% |
| Small signal gain | 32% |
| Optical extraction efficiency | 0.48% |
| Optical Mode | $TM_{00}$ |
| Optical divergence | 2mrad |

## (1.2)   FIR FEL of the Institute of Atomic Energy [4]

In the Institute of Atomic Energy (IAE), a L-band high brightness injector has been built and tested. It is planned to add some $3\pi/4$ mode high gradient accelerator section to make it a driver of a FIR-FEL.

The injector consists of a 100kv, 3ns triode electron gun, one quarter wavelength reeentrant coaxial resonator sub-harmonic buncher of 108 MHz and one TW $3\pi/4$ mode buncher of 1300 MHz composed of 9 cavities This particular mode was adopted for both the buncher and the accelerator section because its high BBU threshold that is important for the high current operation of the system.

The performance up to now is given in the following.

| | |
|---|---|
| Injector energy | 1.8 Mev |
| Micropulse current | > 50 A |
| Micropulse width | ~40 ps |
| Normalized emittance | 0.02 cm-rad |

The Schamatic diagram of the system is shown in Fig. 4.

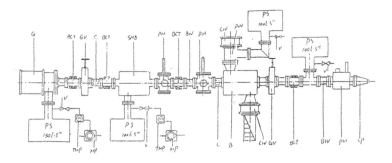

FIGURE 4. Schematic Layout of FIR-FEL of IAE

**(1.3) Wide-band FEL User Facility of the Institute of Nuclear physics and the Institute of Optics and Fine Mechanics.[ 5 ]**

A wide-band FEL user facility (SFEL) is being built at the Shanghai Institute of Nuclear Physics., China Academy of Sciences. The driver of SFEL consists of a triode electron gun, sub-harmonic buncher and a high gradient buncher as injector and Slac type accelerator sections as accelerator. The injector produces micropulses of 59.5 MHz repetition rate of 6-10ps long and 20-100 A current. The designed macropulse width is 6-8 μs. The system layout of SFEL is shown in fig.5.

As can be seen, SFEL uses three different undulators driven by three different energy electron beams to cover a wide wave-length range. The first beam is from the injector itself with energy of 2-3 Mev and produces FIR radiation with wave-length from 800-2000μm. The second beam is produced after the electrons from the injector is accelerated by one accelerator section to an energy of about 30 Mev. This beam, after passing

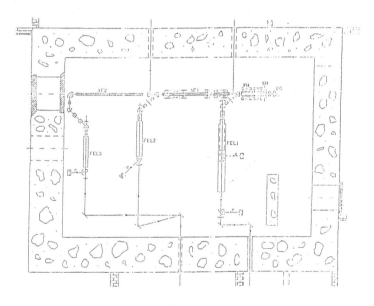

FIGURE 5. Schematic layout of SFEL

through the second undulator, produces Mid-IR radiation from 10-25 µm The third beam, being accelerated.by one more accelerator sections to an energy of 40-50 Mev produces radiation of 2.5-8µm. The wave-length range is designed to cover the applications of FEL in life science, material science and bio-medical science. The use of rf grid controlled triode electron gun and buncher combination allows easy adjustment of bunch separation which is of concern for time resolved experiments.

The above mentioned scheme is for the first phase of the project to further extend the wave-length range to UV, three more accelerator sections and a storage ring will be added as the seciond phase.

The present status of SFEL is that the accelerator sections are available. Buncher is under measurement, gun and modulator being delivered and other components under construction.

### (1.4)  FIR FEL of  China Academy of Engineering Physics [6]

An FIR-FEL facility using a thermionic microwave electron gun as injector operated at 1.3 GHz has been started construction in China Accademy of Engineering Physics. The layout is shown in Fig. 6

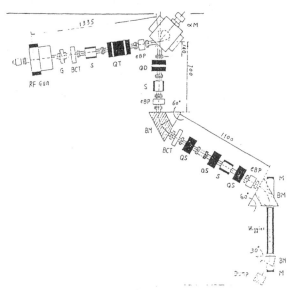

FIGURE 6.  Schematic Layout of FIR-FEL of CAEP

The electron beam from the injector has the following characteristics:

| | |
|---|---|
| Energy | 2.0-2.5 Mev |
| Peak micropulse current | 10A |
| Micropulse width | ~35ps |
| Macropulse current | 450m |
| Macropulse width | 4μs |
| Beam normalized emittance | 15πmm-mrad |
| Beam energy spread | < 1.5% |

## (2)   MM-wave FEL

### (2.1)  Raman FEL Amplifier of China Academy of Engineering Physics (CAEP) [7]

The induction linac based FEL amplifier (SG-1 FEL) of CAEP was started in 1987. Spontaneous emission experiment, constant parameter undulator and variable parameter undulator amplifier experiments were carried out in sequence up to 1994.

SG-1 FEL amplifier is composed of an induction linac, beam transport system, undulator, microwave source ,and diagnotic system as shown in Fig.7.

. The induction linac is composed of a 4-cell injector and a 8-cell accelerator.Each injector cell provides 250kv,100ns pulse voltages to give a total of 1Mv for the diode gun with velvet cathode. The electron beam generated from the gun passing through an axial magnetic guiding field and

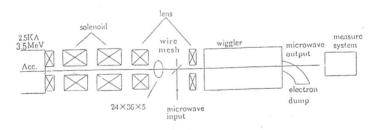

figure 7.  Schematic Layout of SG-1 FEL of CAEP

is injected into the accelerator where each cell gives 300kv acceleration. The measured output electron beam parameters are given by Table 3:

TABLE3.   Beam Characteristics of SG-1 FEL

| | |
|---|---|
| Electron Energy | 3.5Mev |
| Electron Current | 2.5 KA |
| Energy Spreadd | 2% |
| Emittance | 0.43cm-rad |
| Beam Brightness | $10^8$ A/(m.rad)$^2$ |

The accelerated beam passes through beam transport system of solenoid and lens of 2m long and then go through an emittance selector into the undulator. The current there is about 800A.

The undulator is of electro-magnetic type which  has parabolic pole face for double focussing. The focussing strength in y direction is twice that of x direction, so that the equilibrium beam cross-section is elliptical to match the undulator pole shape. The period of the undulator is 11cm and length is 3.96m with nominal field level of 3.1 KGs. Every two period shares one separate power supply so that different tapered field can be produced. Peak magnetic field is continuously variable from 1.4-3.4 kGs on the axis while the two-period at the entrance and at the exit are with lower strength for orbit control. The transmision efficienncy of the undulator is greater than 80% .

Microwave source for SG-1 amplifier is a 20kw, 34.4GHz, 0.35μs pulse-width magnetron. The microwave power entering the undulator region  is about 7kw. Rectangular wave-guide with $TE_{01}$ mode is used for beam-field coupling. The amplified signal frequency and power are measured with dispersion line and crystal detector .

The experimental measurements of the SG-1 system were first performed with the amplification of the spontaneous emission. Power output of 400mw at 100A was obtained which compares reasonably with the result of 600mw according to numerical simulation. Results of the amplification of a seeding signal are illustrated by the following figures. Fig.8 shows the amplifier power output as a function of undulator length for constant parameter undulator and tapered undulator, Fig.9 shows the comparison between numerical simulation and experiments and   Fig.10 shows the output waveforms of  SG-1 FEL experiments.

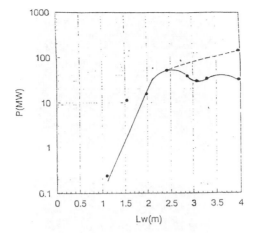

FIGURE 8.  SG-1 Power output  vs undulator length. Dashed line:  tapered undulator.
Solid line:  Constant parameter undulator (3.1KGs).

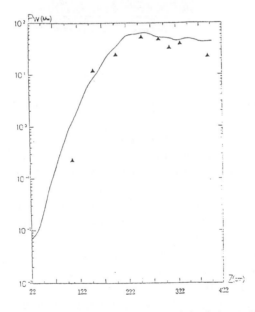

FIGURE 9.  Comparison between experiments (dots) and simulation (solid line)

72    J. XIE

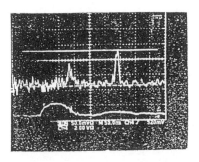

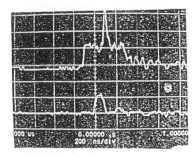

FIGURE 10.    Output waveforms of  SG-1 FEL.  Top: Power output of 140Mw.
Bottom: Upper trace-microwave drive. Lower trace-electron beam.

**(2.2)  Pulse-line Accelerator FEL & Electromagnetic Pumped FEL**
**[8][9]**

FEL with wavelength of several mm has been realized in China since
mid-eighties. Shanghai Institute of Optics and Fine Mechanics first
produced superradiant emission of 8mm wavelength and 1 Mw power with
a 0.5 Mv pulse-line accelerator. South-west Institute of Applied Electronics,
in cooperation with University of Electronic Science and Technology of
China (UESTC), performed similar experiment with an annular beam of 0.7
Mv and radiation of 32 GHz had been observed. Later, amplifier
experiment without axial guiding magnetic field and with a 1.5m, 3.45cm
period undulator and 280A was carried out which produced output power
of 7.6 Mw at 37GHz.[10].

Elecromagnetic pumping FEL was first implemented in UESTC where a
system that produced stimulated radiation with an interaction region

separated from the BWO pump wave generator was used.[11] The radiation generated by the BWO serces as an electromagnetic undulator which has the advantage of producing up-shifted frequency radiation with relatively low source voltage. The schematic diagram of the system is shown in fig.11.

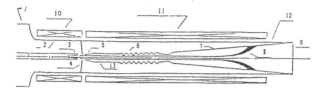

1. pulseline accelerator   2. bellows   3. cathode   4. anode   5. wave cut-off region   6. solw-wave structure for BWO   .7. output horn   8. relativistic electron beam   9. output widow   10. cathode region magnetic-field   11. main magnetic field   12. vacuum cover   13. FEL interaction region

FIGURE 11.   Electromagnetic pumped FEL of UESTC

Electron beam of 0.6 Mev   and 3-5 KA generated by a pulse-line acceleerator passes through the slow-wave structure of BWO at the down-stream and generates microwave of 3cm wavelength as an electromagnetic pump source for the FEL interaction region located at the up-stream. Stimulated radiation of   3-8mm wavelength is produced at the output window and detected with a crystal detector.

**(2.3)  Cherenkov FEL**

CFEL has been pursued in China in several Institutes.[12] Here, only the case of multilayer dielectric loaded waveguide CFEL [13] will be given   as an illustration The calculated longitudinal field distribution with multi-layer dielectric in a plane waveguide shows that the first fundamental mode $TM^1_{01}$ has very favorable field distribution for beam-wave interaction,  The dielectric loss is also found to be lower than the conventional waveguide and the interaction area is increased. Fig.12 (a)  shows the schematic diagram of the CFEL oscillator consisting of five ceramic plates of 445mm long, 0.6mm at top and bottom and 1.2mm for the others in a 26mm*16mm waveguide.driven by a pulsed-line generator. The solid electron beam generated was cut into sheet beams by a metallic screen. About 80% of the generated  beam  entered the cavity.  The useful beam current is 280A at

500KV. The radiation power of 33.4 GHz is in close agreement with that predicted from the dispersion relation as given by Fig.12(b). The output radiation power of 1.7 Mw was measured with a calibrated crystal detector at the end of the dispersive delay line, as shown in Fig.12(c). The corresponding efficiency was about 1.2%.

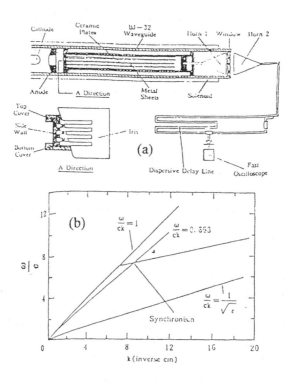

FIFURE 12   Multiple dielectric CFEL of USTD
(a) Schematic diagram
(b) Dispersion relation

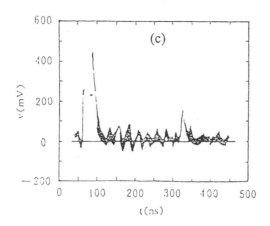

FIGURE 12.  Mutiple Dielectric CFEL of USTD
(c) Output power wave form

**(3) UV FEL**

**(3.1)    FEL Harmonic generator at Hefei National Synchrotron Radiation Laboratory.[14]**
     At Hefei synchroitron Radiation Laboratory, a project has started to build an optical klystron installed in one of the straight section.  A third-harmonic of Nd glass laser of wavelength 3533A is used as input to produce third harmonic of 1178A VUV harmonic coherent radiation   for application research. The schematic layout is shown in the following Fig.13.

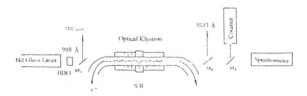

FIGURE 13   Layout diagram of the optical klystron for UV-FEL of Hefei strorage ring.

The electron beam and the optical klystron parameters are given in the following :

| | |
|---|---|
| Energy | 301.6 Mev |
| Energy spread | $1.74 \times 10^{-4}$ |
| Beam current | 50ma |
| Bunch length | 12.3mm |
| Emittance | $2.36*10^{-2}$ mm*mrad |
| Number of bunches | 45 |
| Rotating frequency | 4.8 MHz |
| Energy damping time | 0.195 s |

For the optical klystron:

| | |
|---|---|
| Period | 7.2cm |
| No. of period | 2*12 |
| Field strength | 0.3247 T |
| Gap | 3 cm |
| Length of dispersion section | 21.6 cm |
| Dispersion field strength | 0.7837 T |

It was estimated that with this arrangement, each bunch can produce $1.7*10^5$ photons at 1178A coherently.

## (4) Theoretical Analysis and Numerical Simulation

Besides the construction of experimental facilities as given above, considerable theoretical work on FEL has been accomplished in China. A new analytical method of describing different kinds of Longitudinal modes evolution was proposed. It is a one dimensional, unified theory that includes the effects of pre-bunching, self-bunching and co-bunching caused by strong external field. Co-bunching is responsible for the emission of harmonics and synchrotron sidebands. [15] Linear space-charge wave theory was developed to analyze the bunching process and beam-wave interaction of FEL. [16] Several computer codes have been completed to guide the design and tune-up of the above mentioned facilities. For example, for SG-1, the simulation is made with a 3-D code WAGFEL which, besides other things, includes the space charge effect , transition region of undulator entrance effect etc.[17] This code , after comparing with the experimental data of SG-1 FEL and also with ETA-ELF [18]  proves to be a reliable means to guide the experimental design.


## (5) Components Developments

Peking University has started a super-conducting microwave electron gun project. with Nb cavity operated at 1.5 GHz. The preliminary test of the proto-type showed it can stand accelerating field of 12.6 Mv/m and a Q of $2*10^9$ at 2.1°K.[19]. Fig.14 shows the 3-1/2 cavity thermionic cathode electron gun being developed jointly by Qinghua University and Institute of High Energy Physics.[20]    According to the simulation, the back-bombardment power is only one eighth of the single cavity case through optimization of various parameters.

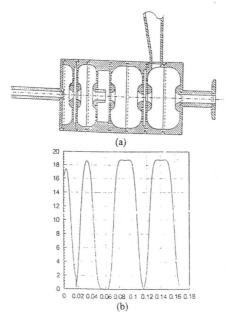

FIGURE 14  A 3-1/2 cavity RF gun under development. (a) Schematic diagram of the cavity, (b) field distribution.

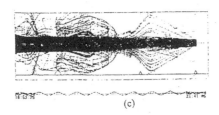

(c)

FIGURE 14   A 3-1/2 cavity RF gun under development. (c) electorn orbit.

### References

[1] X.W. Du, the First Asian Symposium on FEL (1993)
[2] J.L. Xie et.al., Nucl. Instr. & Meth. A341 (1994) 34
[3] V.Kushnir and Gang Wang, Private communication.
[4]  C.G. Yao, IEEE NS32 (1985) 2968
[5] W.Z. Zhou et.al., Nucl. Instr. & Meth. A341 (1994) ABS 37
[6] X.F. Zhao et.al.,the First Asian Symposium on FEL (1993)
[7] K.S. Hu, Private Communication
[8] Z.X. Hui et.al., the First Asian Symposium on FEL (1993)
[9] S.G. Liu, Proc.17th Inter. Conf.on IR & MM  (1992)
[10] K.S. Hu, High Power Laser and Particle Beam, 2 (1990) 158
[11]Z Liang, Proc. of 7th National FEL symposium (1991) 130
[12] C.L. Li, High Power Laser and Particle Beam, 3 (1991) 136; 147
[13] Q.Y. Wang et.al., Appl. Phys. Lett. 59 (1991) 2378
[14] D.H. He et.al., Proc. of 7th National FEL symposium, (1991)
[15] M. Yu, High Power Laser and Particle Beam, 3 (1991) 127
[16] S.G. Liu, Inter. J. Electronics, 72(1992) 161
[17] Z.H. Yang et.al.,High Power Laser and Particle Beams, 1 (1989) 60
[18] T.J. Orzechowski, Proceedings of the Beijing FEL Seminar, (1989) 347
[19] C.E. Chen et.al., the First Asian Symposium on FEL (1993)
[20] C.X. Tang, private Communication

# INFRARED FREE-ELECTRON LASERS: SHORT-PULSE EFFECTS, SPECTRAL PROPERTIES, AND USER ASPECTS

## D. OEPTS

FOM-Instituut voor Plasmafysica 'Rijnhuizen', Edisonbaan 14, 3439 MN Nieuwegein, The Netherlands

**Abstract**    In this contribution we consider some aspects of infrared free-electron lasers in general, and discuss the FELIX facility in particular. The lethargy effect and limit-cycle behavior of short-pulse long-wavelength free-electron lasers are described. It is shown that operation with arbitrarily short electron pulses is in principle possible. Bandwidth and coherence properties are discussed and the possibility of obtaining long-range coherence with short pulses is described. The setup of the FELIX user facility and some of its special features, such as broad-band and rapid tunability (5 to 110 μm wavelength), and flexible bandwidth and pulse format (sub-picosecond to 15 ps transform-limited micropulses) are discussed.

## 1. INTRODUCTION

In a free-electron laser, or FEL, radiation is emitted by electrons that oscillate in an external macroscopic field, instead of by electrons that are bound in an atomic, molecular, or crystalline structure as in other lasers. A relativistic electron beam is injected into the spatially periodic transverse magnetic field of an undulator. The alternating magnetic field experienced by the electrons leads to transverse oscillations and, consequently, to the emission of dipole radiation. As a result of the relativistic speed of the electrons, the emission is concentrated in a narrow forward cone, and the wavelength, $\lambda_0$, is determined by the energy, $\gamma mc^2$, of the electrons, and by the parameters of the undulator, according to the expression

$$\lambda_o = (\lambda_u/2\gamma^2)(1 + K^2), \qquad (1.1)$$

where $\lambda_u$ is the period of the undulator field, K is a dimensionless parameter proportional to the magnetic field amplitude and usually of order unity, and $\gamma^2 \gg 1$ is assumed. In an FEL oscillator, the 'spontaneous' undulator radiation is captured in an optical cavity, and induces newly injected electrons to radiate coherently, which results in the build-up of high laser power.

Although an FEL is similar to other lasers in most respects, there are differences in the basic processes involved, due to the unusual laser medium. In fact, the differences are important enough to sometimes lead to discussions as to whether an FEL is actually a laser. By definition, a laser is a device in which Light is Amplified by Stimulated Emission of Radiation. The FEL conforms to this definition provided that 'light' is understood to include radiation outside the visible range, and that 'stimulated emission' is not exclusively associated with transitions between discrete stationary states. The stimulated coherence of the oscillatory motion of the electrons in an FEL can be explained in a classical model as the result of longitudinal bunching of the electron density in the combined fields of the undulator and the radiation (1-3). Quantum-mechanical corrections are in general small and a full quantum treatment will only be necessary for very short wavelengths or low electron energies (4).

The free-electron laser has the important advantage that it can be made to operate at any wavelength, in principle, by a suitable choice of the parameters in Equation (1.1). Also, a powerful output can be generated, because the electron beam can carry a high energy without the risk of damaging the laser medium or dissipating too much power in the laser structure. In practice, the requirements that the electron beam has to fulfill to obtain laser operation are not easily satisfied, particularly for shorter wavelengths. The electron beams produced by state-of-the-art radio-frequency linear accelerators, or linacs, allow the construction of powerful FELs in the infrared and far-infrared spectral regions. Lasing in the visible and ultraviolet was achieved with electron storage rings. See (5) for a table of existing and proposed FELs .

FELIX, the Free Electron Laser for Infrared eXperiments in Nieuwegein, The Netherlands, operates as a user facility in the wavelength range 5-110 µm. Most of the contents of this chapter is related to FELIX. The hardware and the performance of the installation itself are described in section 4, while sections 2 and 3 are intended to be somewhat more general and discuss short-pulse effects and spectral properties of infrared free-electron lasers.

## 2. SHORT-PULSE EFFECTS

### 2.1 Introduction
Most free-electron lasers operate with an electron beam consisting of micropulses with a duration in the picosecond range, e.g. 1 to 20 ps in the case of radio frequency linear accelerators, or 100-500 ps when a storage ring is used as the electron beam source. The shortness of the electron pulses has an important influence on the operation of the laser. Indeed, the first FEL to operate (6) clearly showed short pulse effects (7).

Whether a pulse is considered short depends on the scale with which the length is measured. The shortest scale that is relevant in the case of a free-electron laser is the wavelength of the radiation. On this scale, the pulses are almost always long, but there are exceptions. Optical pulses with a length of 6 to 10 wavelenghs were observed, and in mm-wave devices, electron pulses shorter than a wavelength are conceivable. These extreme cases will be discussed later in this section.

### 2.2 Slippage
The most important length scale in relation to short-pulse effects is given by the slippage distance, that is, the difference between the distances travelled by an electron and by an optical wave front during their transit through the undulator. In an undulator with $N_u$ periods, the electrons emit $N_u$ wavelengths of radiation at the resonance wavelength $\lambda_0$, and so the slippage distance is equal to $L_s=N_u\lambda_0$. The slippage means that each electron interacts with a portion of the light field of length $L_s$, while each position in the propagating wave field is in turn influenced by the electrons in a length $L_s$ of the electron beam. The effect is illustrated schematically in Figure 1.

When the electron pulse is long compared to $L_s$, there is a large part where the optical pulse develops as if the electron beam were continuous. However, at the beginning and at the end of the pulse there are portions where the optical and electron pulses do not overlap during the whole transit time. The front of the optical pulse interacts with electrons only at the beginning of the undulator, while the trailing edge of the light pulse is influenced by the last electrons just before they leave the undulator. When the electron pulse length is not much longer than the slippage length, the transient effects at the beginning and end of the pulses can become important.

The division between long and short pulses does not always lie just at the slippage length. In a high-gain single-pass device, the light wave reaches a steady state before the end of the undulator, and the full slippage length is

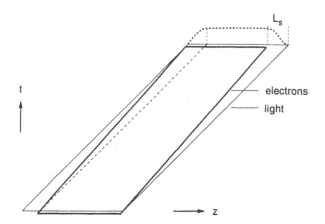

FIGURE 1 Schematic diagram of the propagation of an electron pulse (thick lines) and an optical pulse (thin lines) through the undulator. Transient effects extend over the slippage length $L_s$. The dotted curve at the top indicates the intensity distribution in the light pulse.

not the appropriate length scale to determine whether the electron pulse is short. In that case the electron pulse length should be compared to the cooperation length, which is the slippage in one gain length of the undulator (8). In a low-gain oscillator configuration, the optical pulses are stored in an optical resonator or cavity, formed by mirrors at both sides of the undulator, and are amplified by fresh electron pulses on successive round trips in the cavity. In this case, the end effects accumulate, and become noticeable already for electron pulses that are much longer than the slippage length.

## 2.3 Lethargy

The effect of a short electron pulse on the gain in successive round trips is illustrated qualitatively in Figure 2. In (a) we show an electron pulse and an optical pulse arriving at the entrance of the undulator. On their pass through the undulator, the optical pulse advances by the slippage length with respect to the electron pulse, and at the same time experiences gain (shown exaggerated in the figure). However, the leading part of the optical pulse shows little gain, because it runs away from the electrons already in the first part of the undulator.

In the next round trip, shown in (b), the front of the optical pulse has even a slightly lower intensity due to cavity losses. As a result, in successive round trips, the optical pulse peaks more and more at its end, while it decays at the front. This leads to the phenomenon known as laser lethargy: the optical power grows very slowly, or not at all (7).

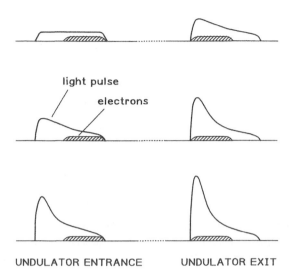

light pulse

electrons

UNDULATOR ENTRANCE          UNDULATOR EXIT

FIGURE 2   Illustration of the effect of slippage on the growth of the optical pulse. The relative positions of an electron pulse and an optical pulse are shown at the entrance and exit of the undulator, in the case that the optical cavity is adjusted for perfect synchronism between the circulating optical pulse and the newly arriving electron pulses. Successive round trips are shown from top to bottom. The leading part of the optical pulse does not grow because it has interaction with electrons only in the first part of the undulator where the electrons have not yet become bunched.

## 2.4  Cavity Desynchronization

The lack of gain at the leading edge of the optical pulse in successive round trips can be compensated by applying a small amount of cavity desynchronization. That is, instead of adjusting the optical cavity length such that the round-trip time for the optical pulse exactly matches the repetition time of the electron pulses, or an integer multiple thereof, the cavity length is decreased by a small amount $\delta L$ from its synchronous value.

The optical pulse then arrives on its next round trip at the entrance of the undulator with a small start relative to the electron pulse, as illustrated in Figure 3. At the exit of the undulator there is initially little difference as compared to the situation in Figure 2, but in further round trips, the part of the optical pulse that had little or no net gain is now in front of the electron pulse. The electron pulse experiences a larger optical field already in the beginning of the undulator, and the optical pulse can grow further without its maximum shifting continually backwards.

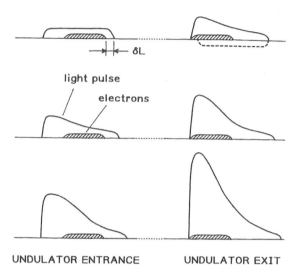

light pulse

electrons

UNDULATOR ENTRANCE            UNDULATOR EXIT

FIGURE 3   Illustration of the effect of cavity desynchronization on the growth of the optical pulse. See text and Figure 2 for details.

Note that we explained in the previous section that the *electrons slip back* with respect to the light wave, while the lethargy resulting from this slippage is now compensated by *advancing the light pulse* even more! This seeming contradiction is due to the fact that the interaction in effect reduces the group velocity of the light pulse. Although the phase velocity of the wave still exceeds the electron speed, the envelope of the optical pulse is changed by the electrons such that its maximum slips back with respect to the electron pulse.

The combination of a reduced group velocity and a shorter optical cavity leads to effective synchronism between the optical and electron pulses, and the optical pulse can develop into a stable shape that grows uniformly. In fact, while the later parts of the pulse grow due to emission by the electrons, the front part is increased just by shifting the larger intensities forward at each round trip.

The stable optical pulse shape that results was described by Dattoli (9) as a *supermode*, that is, a superposition of spatial modes that experience the same gain and phase shift on their passage through the undulator. He also derived a relation between the amount of desynchronization and the effective gain of the supermode. The importance of lethargy is described by the coupling parameter $\mu_c = N_u \lambda_o / \sigma_z$, the ratio between the slippage length and the r.m.s. length of the electron pulse. The long-pulse or small-slippage regime corresponds to $\mu_c \ll 1$. At the optimum desynchronization the small signal gain is found to be given by

$$g = g_0 \frac{1}{(1 + \frac{1}{3}\mu_c)} \, , \qquad (2.1)$$

where $g_0$ is the small-signal gain in the long pulse limit, and the longitudinal density distribution in the electron pulse is assumed to be Gaussian.

The evolution of the optical output power during the macropulse is shown in Figure 4 for different values of the desynchronization $\delta L$ in FELIX. It is seen that the power begins to grow earlier in the pulse for larger values of $|\delta L|$, until at $\delta L = -163$ $\mu$m the desynchronization has been made too large, and the gain has dropped again. As is also seen in Figure 4, the power later in the macropulse is smaller at larger desynchronisms, even though the small-signal gain is higher. This is a result of the different FEL behavior at saturation in a strong optical field.

## 2.5 Saturation

The growth of the optical power due to laser gain is limited by saturation effects. The saturated power level depends on the cavity desynchronization, but not in the same way as the small-signal gain. This is a consequence of the fact that at saturation the gain diminishes and the group velocity of the optical pulse returns to its vacuum value, so that little desynchronism is necessary for optimum overlap between light and electrons.

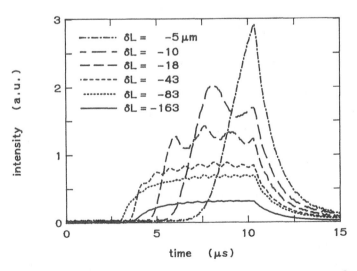

FIGURE 4   Optical macropulses measured in FELIX for various $\delta L$ values at a wavelength of 40 $\mu$m, showing the influence of cavity desynchronization on the initial gain per pass and on the saturated power. The electron pulse ends at 10.5 $\mu$s.

At large desynchronization the optical pulse rapidly begins to outrun the electron pulse when the gain decreases, and stabilizes at a comparatively low power, with a long exponentially shaped front edge. With small desynchronization the power can grow to a higher level, which, however, is reached only at the end of the macropulse, due to the low initial gain. In this case, the optical micropulses are short, because of the peaking up discussed in section 2.3, and the peak power is high. At intermediate $\delta L$ values, the macropulse power is seen to exhibit periodic variations. This effect is discussed in the next section. A combination of high initial gain and high saturated power has been obtained in FELIX by the procedure of dynamic cavity desynchronization, see Section 4.5.2.

At saturation, the strong optical field can lead to complicated nonlinear FEL dynamics. Electrons can become trapped in the ponderomotive potential wells associated with the combination of the optical and the undulator fields, and perform longitudinal oscillations known as synchrotron oscillations, with a period that depends on the optical power density (3). The electrons periodically lose and gain energy to and from the field, depending on the phase of their synchrotron oscillation. The slippage in one synchrotron period is called the synchrotron length, and plays an important role in the evolution of the optical pulses.

One would expect the optical power to reach an equilibrium at such a level that the synchrotron length is twice the slippage length. In that case the electrons perform half a synchrotron cycle during their interaction time in the undulator, and would begin to absorb energy from the field when the field threatens to get larger. This equilibrium is not stable, however, because successive emission and absorption of radiation by the same electrons occurs at different positions in the optical pulse. Locally, the power can still become higher, and a modulation of the optical power with a period of the slippage length can grow unstable in a long pulse (3,10).

This effect is known as the trapped particle instability, or sideband instability because the modulation of the optical power gives rise to sidebands in the spectrum. Usually, the instability leads to irregular, spiking, operation (11). A related but different manifestation of the nonlinear dynamics of the electrons in strong optical fields, known as limit-cycle behavior, is observed in the short-pulse case.

## 2.6  Limit Cycles

In a short-pulse free-electron laser, a dynamical balance between lethargy and saturation can develop, in which the micropulses consist of sets of successively growing and decaying subpulses. This phenomenon of limit-cycle oscillations has long been known from simulations, but has up to now been observed only in FELIX (7,12).

Numerical simulations based on the self-consistent solution of the Maxwell-Lorentz equations for the dynamics of the electrons and the radiation field have been used to study the evolution of the optical micropulses from the small signal to the saturated state (13). The differential equations, in the form given by Colson (3), are solved numerically for a set of sample electrons in successive time steps at a series of sites in the optical pulse. A one-dimensional formulation is commonly used because transverse effects are of secondary importance in this context, although they are not always negligible (14).

Such simulations have shown the development of a stable pulse shape in the linear growth regime, both for small and large cavity desynchronizations. When the gain decreases due to saturation, the desynchronization is no longer compensated by the gain, and the optical pulse envelope begins to advance with respect to the electron pulse. In the case of large desynchronism, the pulse attains a new stationary shape, where gain and loss maintain a stable equilibrium in a smooth pulse that is much longer than the electron pulse. In the case of smaller desynchronism the field can grow to higher power because the overlap between electron and optical pulses is maintained longer. The maximum of the optical pulse slowly shifts forward and continues to grow, until the synchrotron period becomes so short that the electrons begin to absorb at the end of the undulator. As a result, the trailing part of the optical pulse becomes attenuated until its maximum begins to move away from the electron pulse. The optical pulse then loses contact with the electron pulse and advances with a relative speed determined by the desynchronization. At the same time, the suppression of gain by saturation disappears, and a new optical pulse begins to grow. The subsequent growth and decay of subpulses leads to the modulation of the macropulse power seen in Figure 4.

Direct measurements of the optical micropulse length and the formation of multiple pulses in FELIX were made recently (15). Intensity autocorrelation measurements were performed using second harmonic generation in a nonlinear material. A non-collinear background-free setup employing a thin CdTe crystal was used.

Figure 5 shows some results. The average power in the macropulse is shown in Figure 5(a), and measured intensity autocorrelation functions corresponding to different times in the macropulse are depicted in Figures 5(b) to 5(f). The times at which the latter were measured are indicated in Figure 5(a). The wavelength was 24.5 $\mu$m and the desynchronization $\delta L = -16$ $\mu$m. In Figure 5(b) we see the autocorrelation of a smooth pulse with a full width half maximum (FWHM) of somewhat over 800 fs. In Figure 5(c), the three peaks show that the pulse has evolved into

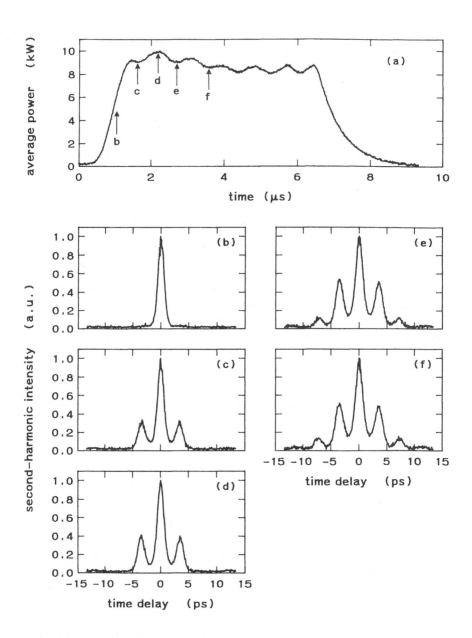

**FIGURE 5**   Intensity autocorrelation traces for five different instants (b-f) in the macropulse illustrated in (a), measured at FELIX, using $\delta L = -16$ μm at a wavelength of 24.5 μm.

two separate pulses, with a separation of 3.5 ps, or 1 mm. The situation in Figure 5(c) corresponds to the moment when the minimum in the first limit-cycle oscillation of the power is reached, the initial optical pulse has lost contact with the electron pulse, and a new pulse is growing. In (d), the second pulse has grown to about the same power as the first, and a maximum in the limit-cycle power oscillation is reached. In (e), a new pulse has formed again, while the initial pulse has decayed but still shows up as a correlation peak at 7 ps. In Figure 5(f), finally, it is seen that a train of four pulses has formed. The presence of the multiple pulses causes sidebands in the spectrum, which were also observed.

Using a smaller desynchronization, it is possible to produce still narrower pulses with higher peak power. Pulse widths of 500 fs FWHM at 24.5 μm wavelength and 220 fs at 10 μm, with a micropulse power of 25 μJ, were observed. These pulse lengths correspond to only six optical wavelengths.

## 2.7 Ultrashort Pulses

### 2.7.1 Electron pulse limit
Before FELIX was built, there has been some doubt as to whether a long-wavelength short-pulse free-electron laser could operate at all, because of the lack of overlap between the optical and electron micropulses. Operation at values of $\mu_c$ up to 10 has now shown to be possible, however.

The question then is, is there a limit to the shortness of the electron pulses, or to the wavelength, beyond which $\mu_c$ is so large that a free-electron laser cannot operate?

When $\mu_c \gg 1$, or $\sigma_z \ll N_u\lambda_0$, then each part of the optical pulse interacts with electrons in only a fraction of the undulator, as illustrated schematically in Figure 6. The reduced overlap between the optical and electron pulses leads to a reduced gain. When we assume that Dattoli's expression, Equation (2.1), still holds in this case, we can approximate the optimum small signal gain by

$$g \approx g_0 \cdot \frac{3}{\mu_c} = g_0 \cdot \frac{3\sigma_z}{N_u\lambda_0} . \tag{2.2}$$

Clearly, this goes to zero when $\sigma_z$ goes to zero. However, this is rather trivial when one realizes that $g_0$ is calculated for long pulses with a fixed current. For such a pulse, the number of electrons in the pulse goes to zero when the pulse length goes to zero, and therefore the gain must vanish as well. To separate the effect of the pulse length from that of the number of

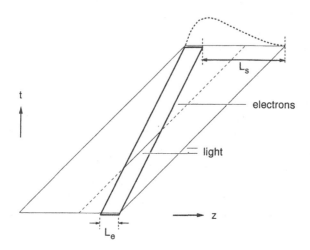

FIGURE 6   Schematic illustration of the overlap between an optical (thin lines) and an electron (thick lines) micropulse in the case of large slippage: $\mu_c \gg 1$, or $L_e \approx \sigma_z \ll L_s$. The dotted line at the top indicates the intensity distribution in the light pulse.

electrons in the pulse, one should determine what happens when the pulse is shortened while the number of electrons is kept constant. Pulses with a fixed number of electrons have a fixed charge, say $q$, and an effective current $I \approx qc/\sigma_z$. Because $g_0$ is proportional to $I$, the simple model (3) then gives a gain proportional to $q$, and independent of $\sigma_z$. Simulations have shown that indeed the small-signal gain, the saturated power, and the optical pulse shape remain essentially constant for $\mu_c \gg 1$ when $q$ is kept constant (16). The proportionality to $q$ does not hold for long pulses ($\mu_c < 1$), where the current still determines the gain. In general, the number of electrons in a slippage length is the appropriate parameter to use.

Does the above mean that there is no short-pulse limit at all? In the low-gain approximation, the electron dynamics is determined only by the incident field and the undulator field; the radiation produced by other electrons is neglected until in the next round trip. This means that there is no communication between electrons in different parts of the same electron pulse, and the FEL interaction remains unchanged even in an electron pulse of one wavelength long. Although the peak current goes to infinity when a pulse with constant charge is made shorter and shorter, the low-gain approximation remains valid when it was satisfied for the long pulse, because, as we have seen, the effective gain does not change.

What then happens when the electron pulse is shorter than the wavelength of the radiation? This also does not prevent FEL operation. On the contrary, it means that the beam is prebunched before entering the undulator, and the spontaneous emission is greatly enhanced. The superradiant emission from such a short pulse needs no further amplification. We will return to this phenomenon in section 3.9.

As far as the radiation mechanism in the FEL is considered, we thus find that there is no lower limit to the length of the electron pulses.

The electron pulses cannot be made arbitrarily short for another reason. When one compresses the electrons in a very small space, the Coulomb forces between them will no longer be negligible as in the long-pulse low-gain case. Although the local charge density can be very high before longitudinal space charge effects become important in the FEL, the difficulty of producing short enough electron pulses limits the feasibility of generating single-bunch coherent emission to long wavelengths.

### 2.7.2 Ultrashort light pulses

So far, we discussed the effects of short electron pulses. Now suppose that we have an *optical* pulse that is shorter than the slippage length, while the electron pulse is longer. The situation is sketched in Figure 7. In this case, the electrons interact with the radiation only during a part of their trip through the undulator. Each electron contributes to the optical gain only as if

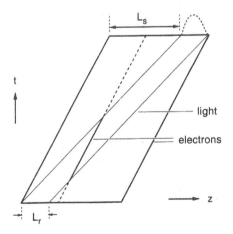

FIGURE 7    Schematic illustration of the overlap between the optical pulse and the electrons in an undulator, in the case that the optical pulse length, $L_r$, is smaller than the slippage length, $L_s$, while the electron pulse is at least as long as $L_s$.

the undulator had its length reduced in the ratio $L_r/L_s$, for an optical pulse of length $L_r$. However, the total gain for the light pulse is larger than it would be for the short undulator, because the light pulse encounters new electrons further down the undulator. In this respect, a single pass of the light pulse is equivalent to multiple passes in an oscillator with a shorter undulator. The short effective undulator length allows a high efficiency, and the pulse can grow to higher power than the normal saturation level.

The situation sketched in Figure 7 will not arise in a small-signal situation, because the bunched electrons will continue to radiate after the incident optical pulse has passed, and the pulse will grow at its trailing edge. Also, with high gain, amplified spontaneous emission can extend the optical pulse at its front edge. In a strong-signal case, however, the emission at the back of the pulse can be suppressed by saturation and re-absorption, and the spontaneous emission can be neglected in comparison with the existing field.

In the case described by Bonifacio (8) as 'strong superradiance', a short and intense optical spike develops at the trailing end of the electron pulse and moves forward by slippage to create a situation as in Figure 7. In general, new optical pulses can develop behind the electron pulse as is the case in spiking and limit cycles. Similar effects occur for ultrashort pulses in other laser media, and in some circumstances a so-called $\pi$ pulse can propagate without changing its shape (17).

Recently, simulations showed that an ultrashort optical pulse can be amplified in a tapered undulator without developing subpulses (18). A 'clean' short optical pulse can also be obtained in the limit cycle regime by a reverse step taper of the undulator (19).

### 2.7.3 The Slowly Varying Envelope Approximation
It is customary to use the Slowly Varying Envelope Approximation, or SVEA, in the theory and simulations of free-electron lasers. That is, the electromagnetic vector potential is written as

$$A(r,t) = A_s(r,t)e^{i(k \cdot r - \omega t)} + c.c \qquad (2.3)$$

and it is assumed that the amplitude varies slowly in comparison to the carrier, i.e., the relative variation of $A_s$ is small over a wavelength and an optical period, so that

$$|\frac{\partial A_s}{\partial z}| << k_z |A_s| \qquad (2.4a)$$

and $\qquad\qquad\qquad |\dfrac{\partial A_s}{\partial t}| << \omega |A_s|.$ $\qquad\qquad\qquad$ (2.4b)

This approximation greatly simplifies the equations, but for optical pulses that are only a few wavelengths long it is clearly not satisfied. Fortunately, it appears that it is not necessary to make the assumptions Equations (2.4) to obtain the same results as were obtained using the SVEA.

To obviate the use of the SVEA, the FEL equations are written in terms of the electric and magnetic field strengths **E** and **B** instead of using **A**. Writing **E** again in the form

$$E(r,t) = E_s(r,t)e^{i(k \cdot r - \omega t)} + c.c \qquad\qquad (2.5)$$

and using $E = -\partial A/\partial t$ (neglecting the scalar potential), one has

$$E_s = -\frac{\partial A_s}{\partial t} + i\omega A_s , \qquad\qquad (2.6)$$

and the term $\partial A_s/\partial t$, which in the SVEA is neglected in comparison with the second term, is now included in $E_s$ and does not appear explicitly in the equations. It has been shown that, under a much weaker condition than Equations (2.4), the resulting equations without using the SVEA are identical to the usual ones obtained with the SVEA, but with **A** replaced by **E** (both in a dimensionless form) (21).

## 3. SPECTRAL PROPERTIES AND COHERENCE

### 3.1 Introduction
The important difference between conventional light sources and lasers is that in a laser the individual oscillators are locked in phase, so that their emitted waves interfere constructively and give rise to a narrow beam of high brightness.

When a laser is switched on, the optical field is initially also weak and incoherent, but a coherent wave builds up while the radiation circulates in an optical resonator or cavity, through the process of stimulated emission. Perhaps even more important than the increased power is the fact that the coherence in the field of a laser extends over much longer times and distances than in the case that the radiators emit independently.

The field in a laser resonator consists of a superposition of cavity modes, that is, waves that fit in the resonator such that the boundary conditions at the end mirrors can be satisfied. Sets of modes with different field profiles transverse to the resonator axis exist, but we will consider only the lowest order ones. The different longitudinal modes are characterized by the number of wavelengths that fit on the cavity round trip length. On the frequency scale, the modes are separated by the mode spacing $\Delta v_c = c/2L_c$, where $L_c$ is the cavity length and c the light velocity. In some lasers, the final field consists of a single cavity mode, and the output has a very long coherence length and a narrow spectral bandwidth.

In this section we will consider some of the properties of free-electron lasers related to bandwidth and coherence.

## 3.2 Spontaneous Emission

As in other lasers, the field in a free-electron laser oscillator starts from spontaneous emission. In the absence of an initial field, the electrons emit spontaneous undulator radiation at the resonance wavelength given by Equation (1.1). The phase of the wave emitted by a particular electron depends on its arrival time in the undulator, hence, on its longitudinal position in the electron beam. If the electron beam were continuous and homogeneous, waves of all phases would be present in equal amounts, and the field would average to zero. In reality, the beam consists of discrete electrons and there is a statistical distribution in the number density of electrons at any position. As a result the field is an incoherent sum over a finite number of contributions, and the resulting field strength is proportional to the square root of the electron density. The power of the spontaneous emission is proportional to the electron density, and the emission can be considered as a result of shot noise in the electron beam.

Each electron emits a wave train of $N_u$ wavelengths, and so the coherence length of the spontaneous emission is equal to the slippage length $L_s$, and the natural bandwidth is $\Delta v_s = c/N_u\lambda_o = v_o/N_u$.

## 3.3 Coherence Growth

The interaction with incident radiation of the appropriate wavelength causes the electrons to become bunched on the scale of the wavelength, so that individual waves add coherently and the power increases. The growth of the power is accompanied by a growth of the coherence length.

The growth of coherence in the optical field takes time; although the stimulated emission is coherent with the incident radiation, this does not imply that there is coherence between the fields at different times. Initially the incident field has only a short range coherence as it originates from spontaneous emission, and the stimulated emission is not much more coherent. In each round trip of the radiation the coherence increases because each electron is influenced by $N_u$ wavelengths of the field, and in turn each position in the wave is influenced by a stretch $L_s$ of electrons. This way, the phase information in the field is spread out, and the coherence diffuses through the wave with a characteristic step size of $L_s$ per round trip.

A simpler explanation of the growth of the coherence is obtained in the frequency domain. The FEL gain has a finite bandwidth $\Delta v_g = v_0/2N_u$, and, assuming that we are in the small-signal regime where the modes are independent, the frequencies close to the peak of the gain profile grow faster than those in the wings. As a result, the spectral bandwidth is reduced by *gain narrowing*.

Is it possible that the width narrows so far that only a single mode remains? Before answering that question we first calculate the number, m, of cavity modes in the bandwidth of the spontaneous radiation: $m = \Delta v_s/\Delta v_c = (v_0/N_u)/(c/2L_c) = 2L_c/\lambda_0 N_u$. As the cavity length cannot be shorter than the undulator length, this is at least equal to $2L_u/\lambda_0 N_u = 2\lambda_u/\lambda_0 = 4\gamma^2/(1+K^2)$, which is a large number, say between 100 and 10000. So, considerable narrowing is necessary to approach the single mode limit. The rate at which the gain narrows can be estimated by expanding the gain profile around its peak. Writing

$$g(v) = g_0 \left(1 - \frac{1}{2}\frac{(v-v_0)^2}{(\Delta v_g)^2}\right) \qquad (3.1)$$

for the fractional gain per pass around the peak at $v = v_0$ and with a width determined by $\Delta v_g$, one finds for the intensity distribution close to $v_0$ after n passes, starting from a flat distribution $I(v) = I_0$:

$$I_n = I_0 e^{ng(v)} = I_0 e^{ng_0} e^{-\frac{1}{2}\frac{(v-v_0)^2}{(\Delta v_g)^2}ng_0}. \qquad (3.2)$$

This has a width $\Delta v_n = \Delta v_g/\sqrt{ng_0}$, and so the spectral width decreases with the square root of the number of round trips, which is quite slow and agrees with the diffusion-like behavior mentioned above. When the gain is higher, the narrowing is faster, but also saturation is reached earlier and the final narrowing at the end of the power growth is still modest. To get an idea, say that the gain narrowing goes on until $e^{ng_0} \approx 10^8$, then we have a narrowing by a factor $\sqrt{8\ln 10} = 4.3$, and the bandwidth will still cover many modes.

When saturation is reached, the laser can show complicated nonlinear behavior in which the modes are no longer independent. In a simple model of homogeneous saturation, the gain is reduced in the same ratio for all modes, and for the highest-gain mode the gain just balances the losses (21). In that case the narrowing of the spectrum continues due to mode competition, but now goes proportional to $1/\sqrt{n\alpha}$, where $\alpha$ is the cavity loss per round trip. This can also be interpreted as: the bandwidth decreases with the root of the number of cavity decay times that have passed (22,23).

In almost all practical cases, it would take an unrealistically long time to narrow the bandwidth down to a single cavity mode spacing. Only for very long wavelengths, where the mode spacing is relatively large, is it conceivable that this bandwidth can be attained. Single-mode operation has been claimed for a far infrared FEL using a high-quality quasi-continuous electron beam (24) but the results have been disputed (25).

## 3.4  Bandwidth Limits

### 3.4.1 Fundamental limit
Suppose that it were possible to have a free-electron laser that actually operated in a single cavity mode, what would then be the smallest possible bandwidth of that mode? In other lasers there is a fundamental limit to the linewidth due to quantum noise. A limiting bandwidth can be found for FELs in a similar way. While the amplitude of the laser field is stabilized by the saturation mechanism, the phase of the field is not fixed. Thermal radiation and the incoherent spontaneous emission that is always present, add phase noise to the laser output which is reflected in the spectral width. The resulting bandwidth depends on the ratio between the noise power and the saturated power. Because the saturated power of a free-electron laser is usually very much higher than the spontaneous power, the limiting bandwidth is very small. The thermal noise appears to be even less important, and it is found that the fundamental bandwidth limit is much smaller than for conventional lasers (26). This is probably only of academic interest, however, as it is not likely that this limit can be attained in practice.

### 3.4.2 Pulsed beams
While a continuous electron beam was implicitly assumed in the above, the minimum bandwidth of any signal is limited by its time duration through the time-bandwidth product $\Delta t \Delta v \geq 1$ (27), and the pulsed operation of most FELs therefore also limits the achievable bandwidth. This limit is known as the transform limited bandwidth.

In the case of storage ring FELs, the electron pulses are relatively long. As was explained by Kim (22), however, the arguments given above for the gain narrowing of the bandwidth can be used analogously in the time domain, leading to pulse narrowing as a result of the time dependence of the gain in the pulsed electron beam. This means that the optical pulse in general becomes shorter than the electron pulse. In a storage ring FEL, saturation is reached due to increasing electron beam energy spread, not because of nonlinear strong field effects. Therefore there is no mode competition in the frequency domain, and at saturation all modes in the gain bandwidth can still oscillate. The finally attainable bandwidth is found to be $\Delta v \approx \sqrt{(\Delta v_g \, \Delta v_e)}$, where $\Delta v_g$ is the gain bandwidth, and $\Delta v_e$ the electron pulse transform limited bandwidth.

For linac-driven FELs in the long-pulse regime, $\mu_c < 1$, the saturation is homogeneous in the frequency domain, so that the spectral narrowing continues at saturation, as mentioned in sect. 3.3. In the time domain the saturation is inhomogeneous, and the temporal width increases again to the electron pulse duration. The limiting spectral bandwidth then becomes the electron pulse transform limited bandwidth (22).

In the short-pulse linac-FEL case, $\mu_c > 1$, the electron pulse transform limit exceeds the gain bandwidth, and there is no simple relation between the spectral width of the radiation and either of these two. The actual optical pulse width depends strongly on the cavity desynchronization applied, and both the pulse length and spectral width can be varied by as much as a factor of ten. The spectral bandwidth was found to be usually the transform limit of the optical pulse length (13).

*3.4.3 Sidebands*

Frequently, the transform limited bandwidth is not attained in linac FELs, because the saturated state does not conform to the simple model of homogeneous saturation. Instead, sidebands can be generated because of the trapped particle instability, or limit cycles involving multiple pulses occur as discussed in section 2. The sidebands can be suppressed by keeping the saturated power low, for instance by using a large desynchronization. Spectral filtering in the cavity can also be applied, except for sidebands that are too close to the carrier (28).

While sidebands increase the spectral width, they have the advantage that they also increase the efficiency and the output power of the FEL. Sometimes the electron dynamics can become stochastic and sidebands of sidebands can develop, leading to a very broad spectrum and very high efficiency (29).

**3.7 Short Pulses, Time Coherence**

Free-electron lasers are in general more useful in applications where high power with a broad spectrum is required, as opposed to cases where long range coherence and narrow-band operation is needed. They are suitable for a different type of coherent phenomena than those involving oscillators with highly stable frequencies.

We can distinguish different types of coherence in optical fields used to probe physical systems. In Figure 8 these are illustrated schematically in the temporal and spectral representations. First, we have the case of incoherent or white light, see Figure 8(a). The pulse length is long compared to the inverse of the spectral width, and the spectrum is broad compared to the inverse of the pulse width. If we had a detector that could show the instantaneous field strength instead of only the power, we would see a noise-like signal. For a system containing many subsystems or degrees of freedom with different resonance frequencies, such a light field could be useful to excite all subsystems simultaneously, for instance to raise the temperature. To obtain any information on the response of the separate systems one would need a more selective source or probe.

The case of narrow-band or quasi-monochromatic radiation is shown in Figure 8(b). The frequency spectrum contains only a narrow range, $\Delta v$, of frequency components. In the time domain, the pulse is at least $1/\Delta v$ long, and our hypothetical field detector would show a sinusoidal signal with a well defined frequency. This signal is coherent over the distance $1/\Delta v$, i.e. if the phase is given at $t=t_0$, it can be predicted everywhere in a range $\Delta t \approx 1/\Delta v$ around $t_0$. Such a field with *phase coherence* is useful, particularly when it is tunable, to excite or probe systems with a specific frequency.

In Figure 8(c) we have the opposite of the previous case, a light field that is confined to a narrow interval in time. The frequency spectrum is at

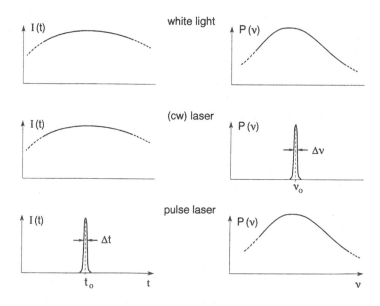

FIGURE 8   Pulse shapes and spectra for optical beams with different temporal coherence properties: incoherent radiation (a), narrow-band phase-coherent radiation (b), and short-pulse time-coherent radiation (c).

least $1/\Delta t$ wide. If one could see not only the power in the spectrum, but also the phase of each component, one would see that these are not random as in case (a), but correlated over a range $1/\Delta t$. Such a field with *time coherence* is useful to study fast dynamical processes in which many frequencies appear. It can be used for instance to excite and probe systems of oscillators with a spread in frequencies, such as molecules in a liquid or solid with different environments influencing their oscillation frequencies. A short but strong optical pulse can coherently excite such an ensemble, i.e. force the phases of the oscillators to be correlated at one time. Subsequent dephasing of the oscillators due to the difference in frequencies can then be studied. A suitable second pulse can reverse the effect of the dephasing, and give rise to a photon echo signal. Short-pulse coherent spectroscopy is probably a more active field nowadays than conventional frequency-domain spectroscopy. Short pulses with high peak power are also useful to study nonlinear optical effects such as multiphoton absorption or frequency conversion.

### 3.8 Long Range Coherence with Short Pulses

In a train of pulses, one can distinguish short and long-range coherence. The coherence length of the separate pulses is, of course, limited by their length, but successive pulses can show coherence between each other on a much longer scale. This is the case in the pulse train from a mode-locked laser. Such a laser uses a medium with a broad gain bandwidth, $\Delta v_g$, so that many cavity modes can be excited. In general, the relative phases of these modes would be independent and variable, and the optical field would be coherent only over a time of the order $1/\Delta v_g$. When mode locking occurs either spontaneously or induced by suitable means, the relative phases of the modes are fixed, and the optical power is concentrated in a short pulse that travels back and forth between the cavity mirrors (21). A fraction of the pulse power is coupled out at each round trip, and the output beam consists of a train of pulses that are mutually coherent.

The situation is sketched in Figure 9, and combines the features of cases (b) and (c) in Figure 8. Both the overall spectrum and the total pulse are broad, but they also contain narrow peaks. The phase locking of the modes in the spectrum makes the pulses short in time, as in 5(c), and the phase coherence between pulses gives the narrow lines in the spectrum, as in 5(b). Note that in this case $\Delta v \Delta t \ll 1$ when $\Delta v$ is the width of one mode, and $\Delta t$ the width of one pulse, while the transform limit usually applies for $\Delta t$ and $\Delta v_g$.

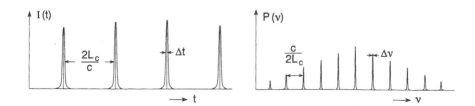

FIGURE 9   Time structure and spectrum of the pulse train from a mode-locked laser. $L_c$ is the cavity length.

Short-pulse free-electron lasers always operate mode-locked due to the pulsed gain, and are normally not very useful for high-resolution narrow-band applications. If narrow-band radiation were desired, one could in principle try to use the long range coherence, and select one of the narrow cavity modes from the spectrum. This is difficult and inefficient because the modes are very closely spaced, and each contains only a very small fraction of the total power.

A mode-locked laser can also operate with more than one optical pulse circulating in the cavity. Without further provisions, these pulses would be independent, and the spectrum would be the same as with a single pulse, though with more power in each mode. However, when the separate pulses are made to be coherent, the spectrum becomes that of a mode-locked laser with a mode spacing corresponding to the distance between the pulses. Communication between the pulses leading to phase locking can be brought about with interferometric means (30). A radio-frequency linac can produce electron micropulses at a high repetition rate to amplify many separate pulses in an FEL cavity.

In FELIX, the electron pulse repetition rate is 1 GHz, and 40 optical pulses are present simultaneously in the cavity with a round-trip time of 25 MHz. Phase locking of the optical pulses by means of a Fox-Smith arrangement causes the suppression of cavity modes that are not multiples of 1 GHz. The phase-locked spectrum contains a much smaller number of modes with a much wider spacing than in the normal set of cavity modes, which makes it possible to select one of the remaining modes with an external filter, and obtain highly monochromatic radiation (31).

The resulting radiation does no longer show a micropulse structure because it contains basically only one cavity mode. The final bandwidth has not been measured yet, but is estimated to be in the order of 1 MHz. The power in the now quasi-cw macropulse amounts to 1-10 W, and so the spectral power density is in the order of 1 W/MHz.

Without the induced phase-locking one should expect no coherence between adjacent optical pulses, because they grow independently from the incoherent spontaneous emission, and there is never communication between them. Each electron pulse has interaction with only one optical pulse on its transit through the undulator, and the distance between the pulses is much larger than their length. Surprisingly, it was found that successive pulses can show a high degree of coherence even without the Fox-Smith arrangement. This effect was found to depend on the settings of the electron accelerator, and can be explained as a result of the phenomenon discussed in the next section.

### 3.9 Spontaneous coherence.

As was mentioned in the introduction, a continuous electron beam leads to incoherent spontaneous undulator emission with a power proportional to the number density of the electrons. As Motz (32) already noted when he studied undulator radiation in 1951, a tremendous increase in power would occur when the electron beam could be produced in pulses shorter than half the wavelength of the radiation. In that case, the electrons would radiate coherently from the start, and the power would be proportional to the *square* of the electron density (1).

The power emitted spontaneously by a bunch of $N$ electrons can be written as

$$P(k) = N P_1(k) + N^2 P_1(k) f(k) \qquad (3.3)$$

where $P_1(k)$ is the power emitted by a single electron at wavenumber $k = 2\pi/\lambda$, and $f(k)$ is a form factor depending on the macroscopic longitudinal density distribution, $S(z)$, of the bunch,

$$f(k) = | \int e^{ikz} S(z) dz |^2 . \qquad (3.4)$$

Here, $S(z)$ is normalized such that $\int S(z) dz = 1$. For a smooth bunch with a characteristic length $l_b$, there will be little constructive interference for wavelengths much smaller than $l_b$. This corresponds to the value of $f$ dropping from unity at $k=0$ to small values for $k > 2\pi/l_b$. However, even when $f(k)$ is very small, the coherent part in Equation. (3.3) can be larger than the incoherent one because of the large additional factor $N$ in the second term. The coherent spontaneous emission can therefore have a noticeable effect on the operation of the laser, specifically on the start-up and coherence properties.

The electron pulses in FELIX, containing $1.2 \times 10^9$ electrons, have a length of about 1 mm (i.e. 3 ps), while the wavelength tuning range is from 5 to 110 μm. Although the pulses are still long compared to even the longest wavelength, effects of coherent emission were observed (33). At a wavelength of 80 μm, a coherent enhancement of the spontaneous emission by a factor of order $10^4$ was observed. At 12 μm wavelength, the enhancement is much smaller, as expected at this short wavelength, but still noticeable.

The behaviour of $f(k)$ for frequencies far from the characteristic $k_b = 2\pi/l_b$ depends strongly on the precise shape of the electron pulse, $S(z)$. The form factor $f$ for two different pulse shapes with the same width of 0.4 mm is plotted in Figure 10. The Gaussian pulse is very smooth on a small scale, and gives an $f$ that decays very rapidly. The other pulse shape has a sharp maximum and a steep edge, which causes coherent effects to extend to much higher frequencies. The shape has been chosen such that the enhancement factor $N.f$ agrees with the observed value ($10^9 \times 10^{-5}$) at 125 cm$^{-1}$ ($\lambda$=80 μm), and such that the k-dependence of $f$, including the oscillatory behaviour, agrees with a qualitative measurement where the resonance wavelength was scanned.

The coherent spontaneous emission also explains the observed coherence between successive pulses mentioned in Section 3.8. The coherent emission is determined by the shape of the electron pulses and has

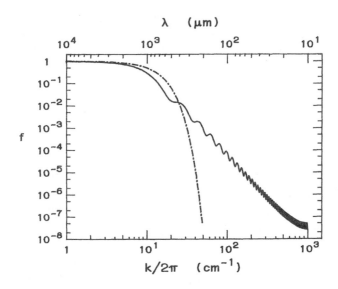

FIGURE 10   Form factor $f$ (Equation (f)) for a Gaussian pulse shape (dotted line) and for a pulse shape assumed for FELIX (full line). Both pulses have width of 0.4 mm (1.2 ps)

a fixed phase relative to these pulses. Consequently, there is a fixed phase relation between the emission from successive pulses. This is in contrast to the incoherent emission, which has a random phase as a result of the independent shot noise variations in each pulse. Interpulse coherence was indeed observed also in the spontaneous signal. The phase locking that is already present in the seed field, is maintained during the process of amplification by stimulated emission. The coherent effects depend strongly on details of the electron pulse shape, and these in turn depend on the accelerator settings, which explains why the latter influence the interpulse coherence. The coherence induced by the finite electron pulses can also be expressed by saying that the FEL radiates at harmonics of the electron pulse repetition frequency. At short wavelengths, very high harmonics are involved, and the effect is weak, but at long wavelengths it becomes more important (34).

At millimeter wavelengths, it is conceivable that electron pulses with a length of the order of the wavelength or even shorter can be used. The spontaneous undulator radiation produced by such a strongly prebunched beam is greatly enhanced by coherent addition. When $f(k) \approx 1$, the spontaneous radiation is fully coherent, and further FEL gain is not possible. This situation is the classical equivalent of spontaneous emission by a coherently prepared ensemble of quantum radiators, known as Dicke superradiance (35). Because of the increased energy loss of the electrons, the spectral properties of the radiation will be different from those of ordinary spontaneous emission. This regime of operation still requires more investigation.

# 4 THE FELIX USER FACILITY

## 4.1 Introduction
FELIX has been designed as a versatile source of radiation in the infrared and far infrared spectral regions. The facility is operated to provide external users with tunable, high power, short pulses of radiation for application in e.g., physics, chemistry and biology. FELIX covers a broad wavelength range and offers an optical pulse structure that can be adapted to the users requirements. It was the first free-electron laser to operate in the short-pulse, large slippage regime, at wavelengths up to 110 μm. Besides serving an international users group, it produces interesting FEL physics results.

In the following sections we give a description of the facility, of its performance and of its special features, and discuss some user aspects.

## 4.2 Basic Machine

To cover the broad wavelength range of 5 to 110 μm, the FELIX machine comprises two lasers working at different electron beam energies. The layout of the installation is diagrammed in Figure 11, the main parameters are summarized in Table 1.

Both FELs are almost identical. The linear undulators consist of two rows of samarium-cobalt permanent magnets forming 38 field periods of 65 mm length. The distance between the rows can be varied to change the K-value of the undulator and thereby scan the wavelength of the laser. Two gold-plated copper mirrors placed at opposite sides of the undulator define the resonator. Behind the undulator the electrons are bent out of the resonator and dumped. The larger part of the space between the mirrors is taken up by electron bending and focussing magnets, etc., rather than by the undulator, which leads to a cavity length of 6 m, with the undulator positioned somewhat asymmetrically at the downstream side as viewed from the electrons. The curvatures of the end mirrors are chosen to give the optical beam a waist roughly at the center of the undulator, and a Rayleigh length of 1.2 m. The mirrors are held in gimbal mounts and the downstream one is carried by a translation stage. Angular as well as longitudinal adjustments can be made with remotely controlled actuators. The distance between the cavity mirrors is controlled to within 0.5 μm by a Hewlett-Packard laser-interferometer system. A fraction of the optical radiation generated in the laser cavity is coupled out through a central hole in the upstream mirror, collimated by a spherical mirror, and transported through

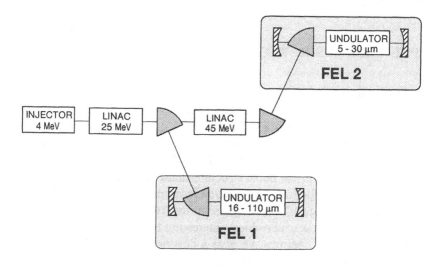

FIGURE 11   Schematic layout of FELIX

TABLE 1  Parameters of the FELIX installation

|  | FEL1 / FEL2 |  |
| --- | --- | --- |
| Wavelength range | 20-110 / 5-35 | μm |
| Spectral width (fwhm) | 0.5-10 | % |
| Micropulse duration (fwhm) | 1-15 / 0.25-8 | ps |
| Micropulse energy (max) | 25 | μJ |
| Micropulse interval (standard) | 1 | ns |
| Micropulse interval (option) | 40 | ns |
| Macropulse duration (max) | 10 | μs |
| Macropulse repetition rate (max) | 5 | Hz |
| Electron beam energy | 14-25 / 25-46 | MeV |
| Energy spread (rms) | 0.25 | % |
| Normalized emittance (rms) | 50 π | mm·mrad |
| Micropulse charge (max) | 200 | pC |
| Undulator period | 65 | mm |
| Number of periods | 38 | |
| RMS undulator strength (max) | 1.3 / 1.9 | |
| Cavity length | 6.0 | m |
| Rayleigh length | 1.2 | m |
| Outcoupling hole diameter | 3 / 2 | mm |

an evacuated beam transport system to the user stations. Diamond windows at the Brewster angle, close to the outcoupling hole, separate the machine vacuum from the optical beam transport tubes.

The electron beam injector is illustrated in Figure 12. A 100 kV thermionic triode gun modulated at 1 GHz delivers 46 mm (280 ps) long electron bunches with a maximum charge of 200 pC. The bunches are compressed to 7 mm in a 1-GHz prebuncher, and further compressed in a 14-cell travelling wave 3-GHz buncher. The rf-power from the buncher exit is fed to the two accelerator sections after appropriate power division and phasing. The bunches are injected into the accelerator with an energy of 3.8 MeV and a length of 0.9 mm. The accelerators are two identical 3-GHz constant-gradient structures, that accelerate the injected electron bunches to

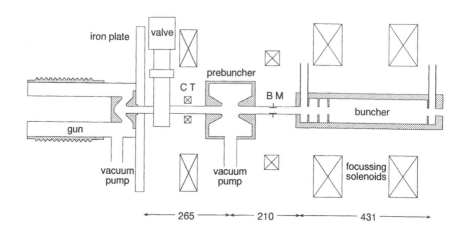

FIGURE 12   Schematic view of the FELIX injector. Dimensions are in mm. CT is a current transformer, BM is a button position monitor.

energies of 14-25 and 25-46 MeV respectively. The energy spread in the electron beam is 0.25% and the emittance $50\pi$ mm·mrad. The electron beams are steered into the undulators by an achromatic bend system (36).

## 4.3   Diagnostics

The electron beam is monitored by means of current transformers behind the injector and at the dumps, and at several positions along the beam line by non-interceptive button position monitors, pop-in fluorescent screens, and leakage current monitors at collimator apertures. An electron energy spectrometer is installed before the dump at the end of FEL2. The spectrometer consists of a bending magnet with a dispersion of 8mm/% in the focal plane. An aluminum optical transition radiation (OTR) target is mounted in this plane at a 45° angle to the beam, and the OTR is focussed on a 32-element photodiode array. This system allows time-resolved measurements of the electron energy spectra both in lasing and non-lasing conditions (37). As an example, Figure 13 shows the evolution of the electron spectrum in the macropulse while the laser operates in the limit-cycle mode. After an initial period of 3 μs in which the energy still varies due to beam loading effects, the energy spread increases due to the energy loss to the optical field. After about 4 μs the limit cycles begin, and both the average electron energy and the energy spread become strongly modulated by the interaction with the optical field.

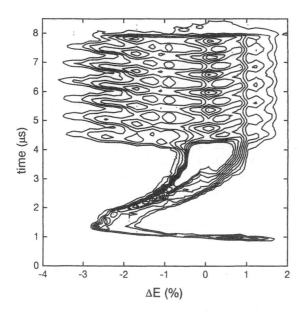

FIGURE 13   Energy spectrum of the electrons emerging from the undulator of FELIX FEL2. The interval between the contour lines is 10% of the maximum signal. The nominal input energy is 36 MeV. Optical wavelength 30 μm, cavity desynchronization -50 μm. Limit-cycle oscillations with a period of 0.6 μs are seen to strongly modulate the electron energy.

Before the optical output beam is sent to the user stations, a fraction of the beam is split off to a diagnostic station for measurement of the macropulse shape and power, and of the optical spectrum. A multichannel optical spectrometer is equipped with a 3-grating turret to cover the full wavelength range, and with a 48-element pyroelectric detector array. The detector elements are read out in parallel, and a time-resolved spectrum can be recorded for each macropulse.

## 4.4 Performance

The electron pulses in FELIX have a spacing of 1 ns and a duration of 3-6 ps. The duration of the pulse train is typically 10 μs, limited to a maximum of 20 μs by power requirements in the accelerator system. Optionally, the micropulse repetition rate can be reduced to 25 MHz. The macropulses are repeated at up to 5 Hz.

The output wavelength of the lasers can be tuned continuously over roughly one octave, on a timescale of minutes, by changing the undulator

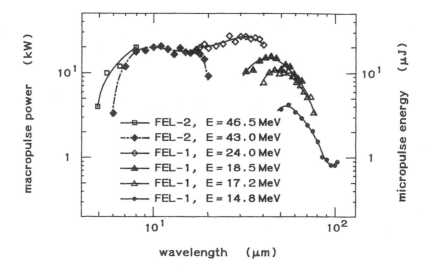

FIGURE 14   Saturated macropulse power and micropulse energy as a function of the radiation wavelength. Results are shown for various values of the electron beam energy. At each electron energy, the wavelength is varied via adjustment of the undulator field.

strength. A larger wavelength change requires a change in the electron beam energy and sometimes a switch to the other laser branch. This takes a time in the order of half an hour, depending on the amount of optimization required and on earlier settings of the accelerator system.

Output powers obtained over the full range of operation are shown in Figure 14. On the left axis the average saturated power on the macropulse timescale is shown, while the right axis gives the energy per micropulse. For a typical micropulse duration of 5 ps, the peak power in the micropulse is in the order of Megawatts. As will be discussed below, the duration of the micropulses and the spectral bandwidth can be varied over roughly an order of magnitude.

The data shown in Figure 14 represent the measured power at the diagnostic station. At each electron beam energy, only the undulator magnet gap was varied. Usually, the output can be increased at a fixed wavelength by optimizing the machine settings for that case. The power available at a user station is somewhat lower due to losses in the beam transport system and output window. The r.m.s. macropulse-to-macropulse stability of the output power is 1-2% when the desynchronization is not too small The relative bandwidth for the cases shown in Figure 14 is in the order of 1%.

## 4.5 Special Features

### 4.5.1 Ultrafast wavelength sweeps

A special feature of free-electron lasers is that the laser medium is refreshed on a nanosecond timescale, and the properties of the medium can be changed on a submicrosecond timescale. This property can be used to change the output wavelength of the laser during a macropulse by changing the electron energy. The electron beam energy is a function of the beam current due to the strong beam loading of the accelerator. The current delivered by the electron gun can be rapidly modulated by changing the bias voltage on the grid, and so the electron energy can be rapidly modulated without changing other accelerator parameters. A total wavelength shift of 2% and sweep rates of 1.4 %/$\mu$s have been realized (38). A macropulse with a changing wavelength could be useful in successively exciting higher levels in an anharmonic molecular potential.

A wavelength variation, or chirp, within a micropulse can also be produced. One way to obtain a chirp is to vary again the electron beam energy and make use of the cavity desynchronization to partly separate the portions of the optical pulse that are generated in different round trips, and therefore have a different wavelength (39).

Another way to produce a chirped pulse is to generate a broad-band short pulse and separate the wavelengths outside the laser with a dispersive device as discussed in the next section.

### 4.5.2 Flexible pulse format and spectrum

As was discussed in section 2.8, variation of the cavity desynchronization allows the generation of either long or short pulses, with a narrow or broad spectrum. A factor of ten variation in micropulse length and in spectral bandwidth can be obtained. Pulse lengths of 0.3 to 3 ps can be produced in the short wavelength range, while bandwidths corresponding to pulses of 1.2 to 14 ps were observed at a wavelength around 40 $\mu$m (13).

Many users want to have a short pulse with high peak power. This requires the use of small desynchronization, which has the drawback that the initial gain is small and the laser operates just above threshold, so that the output is not very stable. Also, the limit-cycle behavior shown at larger desynchronizations, although interesting for FEL physicists, is not convenient for users, who want a stable pulse shape throughout the macropulse. The solution to these problems is to start the laser with a large desynchronization and to switch to a small one at the beginning of saturation. It is not possible to actually change the cavity length at the required time scale, but an equivalent adjustment of the electron micropulse repetition rate is possible.

The effect of Dynamic Cavity Desynchronization is illustrated in Figure 15. Curve (c) shows the macropulse power obtained with an initial desynchronization giving a high small-signal gain, as in curve (a), and a final desynchronization as in curve (b), leading to a high peak power. The transition was made by feeding a suitable ramp signal to the FM input of the master oscillator of the accelerator system, causing a relative frequency change of $4 \times 10^{-6}$ at 4 to 5 µs after the start of the accelerator macropulse. The resulting optical pulse combines high power, short micropulse duration, and improved stability both within the macropulse and from macropulse to macropulse (40).

The optical micropulses can also be manipulated with an external pulse shaper. This device is equivalent to a grating rhomb as commonly used in the visible and near-infrared, but uses only reflective optics and a single grating imaged onto itself with a magnification of -1 (41). The pulse shaper can be used to stretch short pulses into long chirped pulses. For instance, a 0.5 ps pulse at a wavelength around 7.7 µm was stretched to 16.5 ps with a wavelength chirp in the order of 0.3% per picosecond. Also, chirped laser pulses can be compressed.

Some users do not want the high micropulse repetition rate of 1 GHz because of undesired heating of their sample, or because they want to study relaxation times that are not short compared to 1 ns. For these cases, a single

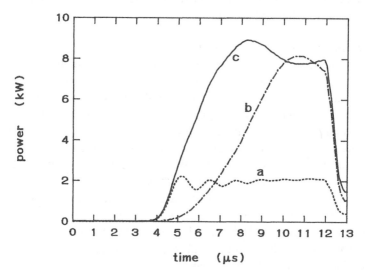

FIGURE 15   Optical macropulse measured at a wavelength 42 µm, (a) using a fixed desynchronization of $\delta L$ =- 27 µm giving a high small-signal gain, (b) using a fixed $\delta L$ =- 4 µm giving a high saturated power, and (c) with dynamic cavity desynchronization in which $\delta L$ is switched from -27 to -4 µm at t≈4 µs.

micropulse can be sliced from the macropulse with the aid of a silicon slab that is switched from a transparent to a reflective state by illumination with the pulse from a Q-switched Nd:Yag laser (42).

As another option to reduce the micropulse repetition rate, an additional modulation can be added to the electron gun such that the electron pulse repetition frequency is reduced to 25 MHz, which is the round-trip frequency of one optical pulse in the cavity.

The wavelength range of the facility is further extended at the short-wavelength side, down to 3 μm, by generation of second-harmonic radiation in GaSe and $ZnGeP_2$ (43).

### 4.6 User Aspects

A free-electron laser is an expensive device, and it is necessary to optimize proper use of the beam time. At the FELIX facility, a Programme Advisory Committee considers user proposals for experiments, and advices on the allocation of beam time on the basis of their scientific merits and the necessity of using FELIX. A Facility Manager discusses with the users the requirements of the experiment and the technical possibilities of the facility beforehand, so that provisions can be made when necessary. FELIX staff assists the users in setting up equipment and, when necessary, make adaptations to instruments or outfit of the assigned user station.

Multiple user stations are accommodated, so that it is not necessary to break down one setup before another experimentalist can begin building up his apparatus. More or less permanent setups are maintained for groups of similar experiments. These are controlled by a station master based at the facility, who assists different users from the same discipline, using his accumulating experience in working with FELIX and the associated instrumentation in his station, such as cryostats, magnets, and data acquisition systems. Short-term or incidental users are encouraged to use one of the existing setups to save time.

A two shifts a day, five days a week schedule is maintained for normal operation. Usually, two different user groups alternate shifts for a few days. Experienced long-term users can operate the machine themselves in the weekends. Two-shift operation started in July 1994 and in that year 2200 hours of operation were made available to users. The target for a full year of operation in this mode is 2800 hours.

The output of the lasers is guided to the different user stations by means of a beam tube system with pneumatically activated switching mirrors. The beam tubes are evacuated to avoid absorption of the radiation by atmospheric gases, particularly water vapor. The main user area is shown in Figure 16. Each user station contains one or more optical tables, and basic utilities like

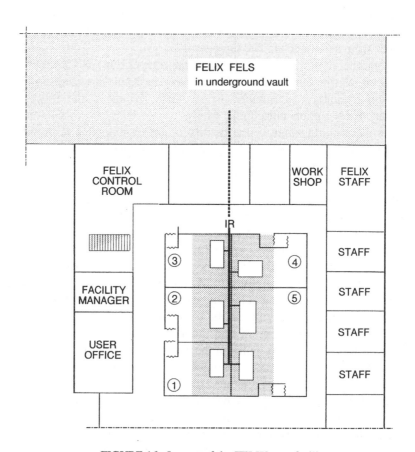

FIGURE 16   Layout of the FELIX user facility

electricity, cooling water, dry nitrogen gas, and roughing vacuum, in addition to a connection to the optical beam tube distributing the FELIX output, and trigger signals for synchronization purposes. Three more user stations, and also the diagnostic station where the spectrum and macropulse power are routinely monitored, are located in the basement under the main user area. In a later stage, the diagnostic station is to be extended so that operations on the optical beam such as attenuation, filtering, or pulse slicing can be performed before sending the beam to the users.

## 4.7  Applications

Soon after first lasing of FELIX FEL1 in August 1991, exploratory measurements were performed by external users (44). Operation of FEL2 followed in August 1992, and enabled users to start experiments also at shorter wavelengths (45). Routine operation as a user facility started in 1993 after the equipment of the user area infrastructure.

A number of user groups have applied FELIX in different fields of research since 1993. Figure 17 shows the distribution over research areas in 1993-94. The largest fraction is taken up by solid-state physics, where many characteristic frequencies are within the wavelength range of FELIX. The high peak power of the pulses enables the study of nonlinear effects, e.g. in GaAs/AlGaAs (46), while the short duration of the pulses allows time resolved measurements of, e.g., free-induction decay of shallow-donor transitions in GaAs:Si (47). The tunability, flexible pulse shape, and high power are particularly useful features in selecting optimum conditions for biomedical applications of lasers in, e.g., ablation of eye tissue and hard dental material (48).

The requests for beam time exceeded the amount available on all occasions for the submittal of proposals so far. The fraction of projects from outside the Netherlands has steadily increased; for the first half of 1995, 44% of the beam time has been allocated to research groups from six different countries: Belgium, Germany, Russia, Switzerland, the UK and the USA. This shows that FELIX operates as an international facility offering new research possibilities to scientists in a variety of disciplines.

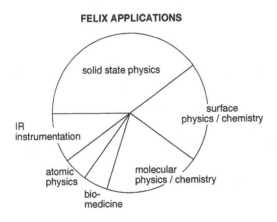

FIGURE 17   Distribution of FELIX applications over different fields of research

## ACKNOWLEDGMENTS

This work was performed as part of the research programme of the 'Stichting voor Fundamenteel Onderzoek der Materie' (FOM) with financial support from the 'Nederlandse Organisatie voor Wetenschappelijk Onderzoek' (NWO).

## REFERENCES

1. C. Pellegrini, *1-D theory of high gain FELs*, this volume.
2. T.C. Marshall, *Free Electron Lasers*, (MacMillan, New York, 1985).
3. W.B. Colson, in *Laser Handbook*, Edited by W.B. Colson, C. Pellegrini, and A. Renieri, (North Holland, Amsterdam, 1990), Vol.6 p.115.
4. G. Dattoli and A. Renieri, in *Laser Handbook*, Edited by W.B. Colson, C. Pellegrini, and A. Renieri, (North Holland, Amsterdam, 1990) Vol.6 p.221.
5. W.B. Colson, *Nucl. Instrum. Meth.* A **358**, 555 (1995).
6. D.A.G. Deacon, L.R. Elias, J.M.J. Madey, G.J. Ramian, H.A. Schwettman, and T.I. Smith, *Phys. Rev. Lett.* **38**, 892 (1977).
7. H. Al-Abawi, F.A. Hopf, G.T. Moore, and M.O. Scully, *Opt. Commun.* **30**, 235 (1979).
8. R. Bonifacio, B.W.J. McNeil, and P. Pierini, *Phys. Rev. A* **40**, 4467 (1989).
9. (a) G. Dattoli, A. Marino and A. Renieri, *Opt. Commun.* **35**, 407 (1980).
   (b) G. Dattoli, and A. Renieri, in *Laser Handbook*, Edited by M.L. Stitch and M. Bass, (North Holland, Amsterdam, 1985).Vol.4 p.1.
10. N.M. Kroll, P.L. Morton, and M.N. Rosenbluth, *IEEE J. Quantum Electron.* **QE-17**, 1436 (1981).
11. R.W. Warren, J.E. Sollid, D.W. Feldman, W.E. Stein, W.J. Johnson, A.H. Lumpkin, and J.C. Goldstein, *Nucl. Instrum. Meth.* A **285**, 1 (1989).
12. D.A. Jaroszynski, R.J. Bakker, A.F.G. van der Meer, D. Oepts, and P.W. van Amersfoort, *Phys. Rev. Lett.* **70**, 3412 (1993).
13. R.J. Bakker, D.A. Jaroszynski, A.F.G. van der Meer, D. Oepts, and P.W. van Amersfoort, *IEEE J. Quantum Electron.* **QE-30**, 1635 (1994).
14. (a) G.H.C. van Werkhoven, B. Faatz, and G.M.H. Knippels, *Opt. Commun.* **118**, 551 (1995).
   (b) Zili Weng and Yijin Shi, *Nucl. Instrum. Meth.* A **359**, 610 (1995).
15. G.M.H. Knippels, R.F.X.A.M. Mols, A.F.G. van der Meer, D. Oepts, and P.W. van Amersfoort, *Phys. Rev. Lett.* **75**, 1775, (1995).
16. (a) D.A. Jaroszynski, D. Oepts, A.F.G. van der Meer, P.W. van Amersfoort, and W.B. Colson, *Nucl. Instrum. Meth.* A **296**, 480 (1990).
   (b) J. Blau, R.K. Wong, and W.B. Colson, *Nucl. Instrum. Meth.* A **358**, 441 (1995).

17. A. Icsevgi and W.E. Lamb, Jr, *Phys. Rev.* **185**, 517 (1969).
18. T.B. Zhang and T.C. Marshall, *Phys. Rev. Lett.* **74**, 916 (1995).
19. D.A. Jaroszynski, R. Prazeres, F. Glotin, J.M. Ortega, D. Oepts, A.F.G. van der Meer, G.M.H. Knippels, and P.W. van Amersfoort, *Phys. Rev. Lett.* **74**, 2224 (1995).
20. (a) E.H. Haselhoff, *Phys. Rev. E* **49**, R47 (1994).
    (b) W.B. Colson, *Nucl. Instrum. Meth.* **A 341**, ABS64 (1994).
21. A.Yariv, *Introduction to Optical Electronics*, (Holt, Rinehart and Winston, New York, 1971) Ch. 6.
22. K.-J. Kim, *Nucl. Instrum. Meth.* **A 304**, 4458 (1991).
23. T.M. Antonsen, Jr. and B. Levush, *Phys. Fluids B* **1** ,1097 (1989).
24. L.R. Elias, G. Ramian, J. Hu, and A. Avnir, *Phys. Rev. Lett.* **57**, 424 (1986).
25. T.M. Antonsen, Jr. and B. Levush, *Phys. Rev. Lett.* **62**, 1488 (1989).
26. A. Gover, A. Amir, and L.R. Elias, *Phys. Rev. A* **35**, 164, 1987.
27. We use this form for convenience instead of $\Delta t \Delta \nu \geq 1/4\pi$ which holds for the root-mean-square widths, by assuming a modified definition for $\Delta t$ and $\Delta \nu$.
28. S. Riyopoulos, *Phys. Plasmas* **1**, 3078 (1994).
29. P. Chaix, D. Iracane, and C. Benoist, *Phys. Rev E* **48**, R3259 (1993).
30. D. Oepts and W.B. Colson, *IEEE J. Quantum. Electron.* **QE-26**, 723 (1990) .
31. D. Oepts, A.F.G. van der Meer, R.J. Bakker, and P.W. van Amersfoort, *Phys. Rev. Lett.* **70**, 3255 (1993).
32. (a) H. Motz, *J. Appl. Phys.* **22**, 527 (1951).
    (b) H. Motz, W. Thon, and R.N. Whitehurst, *J. Appl. Phys.* **24**, 826 (1953).
33. D.A. Jaroszynski, R.J. Bakker, A.F.G. van der Meer, D. Oepts, and P.W. van Amersfoort, *Phys. Rev. Lett.* **71**, 3798 (1993).
34. A. Doria, R. Bartolini, J. Feinstein, G.P. Gallerano, and R.H. Pantell, *IEEE J. Quantum Electron.* **QE-29**, 1428 (1993).
35. (a) R.H. Dicke, *Phys. Rev.* **93**, 99 (1954).
    (b) A.E. Siegman, *Lasers*, (University Science Books, Mill Valley, 1986), Sect. 13.8
36. C.A.J. van der Geer, R.J. Bakker, A.F.G. van der Meer, P.W. van Amersfoort, W.A. Gillespie, P.F. Martin, and G. Saxon, *Proc. 3rd Eur. Particle Conf., Berlin 1992*, (Editions Frontière, Gif-sur-Yvette, 1992), p. 504.
37. W.A. Gillespie, A.M. MacLeod, P.F. Martin, G.M.H. Knippels, A.F.G. van der Meer, E.H. Haselhoff, and P.W. van Amersfoort, *Nucl. Instrum. Meth.* **A 358**, 232 (1995).
38. G.M.H. Knippels, A.F.G. van der Meer, R.F.X.A.M. Mols, D. Oepts, and P.W. van Amersfoort, *Infrared Phys. Technol.* in press, (1995).
39. G.M.H. Knippels, A.F.G. van der Meer, R.F.X.A.M. Mols, D. Oepts, and P.W. van Amersfoort, *Proc. 17th Int. Free Electron Laser Conf., New York, 1995*, to be published (North-Holland, 1996).
40. R.J. Bakker, G.M.H. Knippels, A.F.G. van der Meer, D.Oepts, D.A. Jaroszynski, and P.W. van Amersfoort, *Phys Rev E* **48**, R3256 (1993).

41. G.M.H. Knippels, A.F.G. van der Meer, R.F.X.A.M. Mols, P.W. van Amersfoort, R.B. Vrijen, D.J. Maas, and L.D. Noordam, *Opt. Commun.* **118**, 546 (1995).
42. E.H. Haselhoff, G.M.H. Knippels, and P.W. van Amersfoort, *Nucl. Instrum. Meth.* A **358**, ABS 28 (1995).
43. J.M. Auerhammer, A.F.G. van der Meer, P.W. van Amersfoort, Q.H.F. Vrehen, and E.R. Eliel, *Opt. Commun.* **118**, 85 (1995).
44. (a) J. Burghoorn, V.F. Anderegg, T.O. Klaassen, W.Th Wenckebach, R.J. Bakker, A.F.G. van der Meer, D. Oepts and P.W. van Amersfoort, *Appl. Phys. Lett.* **61**, 2320 (1992).
    (b) M.F. Kimmitt, C.R. Pidgeon, D.A. Jaroszynski, R.J. Bakker, A.F.G. van der Meer, and D. Oepts, *Int. J. Infrared and mm Waves* **13**, 1065 (1992).
45. M. Barmentlo, G.W. 't Hooft, E.R. Eliel, E.W.M. van der Ham, Q.H.F. Vrehen, A.F.G. van der Meer, and P.W. van Amersfoort, *Phys. Rev. A* **50**, R14 (1994).
46 M. Helm, T. Fromherz, B.N. Murdin, C.R. Pidgeon, K.K. Geerinck, N.J. Hovenier, W.Th. Wenckebach, A.F.G. van der Meer, and P.W. van Amersfoort, *Appl. Phys. Lett.* **63**, 3315 (1993).
47. P.C.M. Planken, P.C. van Son, J.N. Hovenier, T.O. Klaassen, W.Th. Wenckebach, B.N. Murdin, and G.M.H. Knippels, *Phys. Rev. B* **51**, 9643 (1995).
48. B. Jean, R. Walker, T. Oltrup, T. Bende, and P.W. van Amersfoort, *Ophtalmol.* **35**, 2155 (1994).

# NEW INTENSE ELECTRON BEAM GENERATION BY PSEUDOSPARK DISCHARGE*

Ming Chang Wang, Junbiao Zhu, Zhijiang Wang, Lifen Zhang, Yu Huang, Chengshi Feng, Bin Lu, Jae Koo Lee**

Shanghai Institute of Optics and Fine Mechanics, Academia Sinica, P.O.Box 800211, Shanghai, P.R. China.
** Pohang Institute of Science and Technology, P.O. Box 125, Pohang, 790-600, Korea

**Abstract** A brief review of the activities of free electron laser at SIOFM is given. New approach to high brightness beam for FEL is investigated. The design and construction of the pseudospark discharge is described. A high power, high current density and high voltage electron beam was generated with the pseudospark discharge, the experiments are presented. The intense electron beams with the voltage of 200 KeV, current of 2 KA, beam diameter of less than 1 mm and beam emittance of 48 mm mrad were obtained. A compact free electron laser with a table size is discussed.

* Work supported by the National Natural Science Foundation of China.

117

## 1. Free electron laser activities at SIOFM

The first Raman free electron laser in China was lasing in 1985 in our institute, corresponding to laser wavelength of 8 millimeter and peak power of 1 megawatt [1]. In recent years, several new technologies of free electron lasers, such as optical klystron, distributed feedback cavity and small-period wiggler, have been successfully investigated and developed.

A wiggler is an important element for free electron lasers(FEL's), which couples the energy of electron beams to optical wave. In order to achieve the shortest possible wavelength with given electron energies, the wigglers with small periods have to be considered.

Using small-period wiggler is an important method to decrease the FEL wavelength and to tune the laser frequency. We constructed a novel small-period wiggler, which consists of bifilar-helix sheets with ferro-magnetic materials. The configuration of new wiggler is showed in Fig.1. It is similar to the bifilar helical wires wiggler, but in this configuration, there are the copper sheets instead of the wires. In order to enhance the wiggler fields, the conducting sheets of copper are wound through a stack of helical ferro-magnetic cores in alternating directions. This type of wiggler produces circular polarized fields. The maximum transverse wiggler field as high as 1500 G has been obtained with the wiggler period of 10 mm and the gap of 16 mm. The experimental results indicated that the FEL operated in the wavelength of 3 millimeter [2].

There are two difficulties in free electron laser oscillators: the cavities impede the electron beam passing through and the high laser power will damage the cavity mirrors. In order to overcome these difficulties, we adapted a distributed feedback cavity with sinusoidal corrugation, which was used to select the operating mode. Instead of the mirrors, the inner wall of the waveguide consists of two corrugation sections. If the corrugation has a period satisfying the Bragg condition, the longitudinal wave will have feedback on the wall.

The experimental results of Raman FEL oscillator showed that the FEL spectrum became narrow and peak power enhanced [3]. The range of laser frequency become narrow, and the FWHM of pulse is from 7.5 mm to 9.2 mm in superradiation amplifier and from 8.6 mm to 8.9 mm in distributed feedback oscillator. The Fig. 2 shows spectrum measurement of the FEL oscillator with the Bragg cavity.

## 2. Design of the pseudospark discharge(PS)

The high quality electron beams can be produced by the pseudospark discharge, there are many advantages. The beams have intense current density ($> 10^4$ A/cm$^2$), narrow beam diameter ($< 1$ mm) and very low emittance(tens of mm mrad), high brightness ($10^{12}$ A/m$^2$rad$^2$)[4].

We have designed the device and collaborated the research work with POSTECH since 1993.

It is well-known that the electrical breakdown in gases is described by Paschen law $U = f(Pd)$, where P is the gas pressure and d is the distance between the cathode and anode(K-A). A typical breakdown curve as a function of Pd is shown in Fig. 3. It is characterized by near linear rise at pd-values from 25 mbar mm to 40 mbar mm. A minimum around 7 to 13 mbar mm and a steep rise below the minimum. The breakdown below $10^4$ mbar mm is called vacuum breakdown [5].

The condition of the hollow cathode(HC) effect can be written approximately as follows:

$$\varphi P < 133.3 \text{ cm Pa} \qquad (1)$$

where φ is the internal diameter of the hollow cathode cylinder.

The cutoff voltage of single gap PS chamber is limited about 20 KV due to local micro discharges resulting from gas adsorption on the surface of an insulator and local field enhancement at triple points. It is necessary to consider a multi-gap geometry in order to design a high voltage (higher than 100 KV) PS source of electron beam. Generally speaking, the relation between the breakdown voltage $U_b$ and the product of the K-A distance d and pressure P filled in the chamber obeys the Paschen experimental law, $U_b = f(Pd)$. It is experimentally shown that the PS breakdown relation is a function of $P^2d$, not Pd [6]. The relation is as follows:

$$U_b(P, d) = \frac{\alpha}{(P^\beta d)^\delta} \qquad (2)$$

where $\alpha = 0.1865 \pm 0.0019$, $\beta = 1.9952 \pm 0.0064$, $\delta = 2.226 \pm 0.016$, it is seen that ß is approximately 2. With the help of numerical simulation, we find the relation between the K-A distance d and the pressure P filled in the multi-gap geometry at fixed voltages 100-500 KV applied by our modified pulse-forming line(PFL) as shown in Fig. 4. It follows from Fig. 4 that the K-A distance increases with the pressure value P decreasing at a constant voltage; the K-A distance drops with the voltages increasing at a given gas pressure; In the range of 100-500 KV, the curves at higher pressure tend to approximately 2 cm, while at lower end of pressure, reasonable K-A distance should be 10 cm. Therefore the K-A distance is determined between 2 cm and 10 cm depending on applied voltages and filled gas pressure, while the corresponding values of pressure being from 20 pa to 7 pa. By comparison with Eq.(1), this range meets the condition for the HC effect.

## 3. Configuration of the pseudospark discharge

The schematic of the general experimental design of the PS chamber as an intense current density, low emittance, high brightness electron beam source is shown in Fig. 5. The design of the intermediate electrodes and insulators of the pseudospark chamber is similar to W.Bauer's one, which has been proven a high voltage working device [7]. The configuration parameters of the modified pulse line accelerator(PLA) are as follows.

The PS hollow cavity has a 3 cm diameter and 4.1 cm long. The discharge chamber consists of a planar cathode with hollow cavity, sets of intermediate electrodes and insulators with a common channel, and a planar anode. The electrodes are made of brass and the insulators are made of Plexiglas. The diameter of the channel is 3.2 mm. The anode-cathode gap distance is varied in 10-100 mm. A Rogowski coil, a Faraday cup was installed in the drift tube.

A 10-order Marx generator is charged to a higher pulse voltage of more than 100 KV, which has parallel charge and series discharge. After the Marx generator being triggered, the voltage is directly applied to the multi-gap PS chamber via the modified PFL. The PFL was modified to use as a water coaxial cable, by directly shorting its main switch and taking off the ground inductance. No any matching resistance was connected between the modified PFL and the PS chamber due to this low resistance and high power device. Typical pulse voltage waveform applied over the PS chamber, i.e, the output of the PLA, is shown in Fig. 6, from which it follows that the voltage waveform is a rectangular one, whose longer pulse duration and steep pulse front meet the requirements of pseudosparks. The HC effect takes place in the HC operated at low pressure due to the injection of positive ions and the electron ionization and avalanche in it,   with the explosive electron emission near the cathode hole,   resulting in the   step-by-step breakdown of the multi-gap PS chamber from the cathode to the anode. The protecting sheath of each insulator is set up near the discharge channel to screen various radiations due to ionization and bombardments   from runaway particles, therefore effectively avoiding local micro discharges.   A high power, intense current electron beam   is ejected out of the anode hole due to electrostatic   focusing and acceleration.

## 4. Experimental results and further work

High power,   high current density and high voltage electron beam was generated with pseudospark discharge.   The electron beams have voltage of 200 KeV,   current of 2 KA and beam diameter of less than 1 mm.   The beam   penetrated   a 0.3 mm hole on a copper foil of 0.05 mm thick at the distance   of 5 cm from the anode and penetrated   a 0.6 mm hole on a radiachromic film at the distance of 15 cm [8].

The typical waveform of the PS chamber voltage photographed   with an OK-19 oscilloscope is shown in Fig. 6.   It is clear that the voltage pulse approximately is a rectangular wave with a maximum peak of 200 KV,   the pulse duration about 3 $\mu$s and   a rapid rising wavefront.

It is found that the emittance is sensitive to the gas pressure.   The gas pressure of the chamber is controlled   by a Vacuum Auto Controller,   with measuring and control range from 0.15 to $7.5 \times 10^7$ Torr.   The Nitrogen gas is filled by a piezo-electric needle valve, with the pressure accuracy of $\pm 1.5\%$ (full scale) and the maximum flow $> 2.25 \times 10^3$ Torr cm$^3$/s.

Fig. 7 shows the optical density distribution   on the radiachromic film. The measured rms(root-mean-square)   emittance of the electron beam was found to be $\epsilon \sim 48 \pm 10$ mm mrad.    The beam normalized    rms emittance was $\epsilon_n \sim 47 \pm 10$ mm mrad.    It is much less than before.

The beam current will be enhanced and the voltage will go to 300 KV. The beam brightness of the PS is expected to be comparable to a photo-cathode source, the PS beam source has low cost and is easy to fabricate. The design of a compact FEL with a table size is proposed [9]. The original device of a Raman FEL is shown in Fig. 8. The Marx generator can be reduced and the PFL can be omitted if we use PS. The size of a compact FEL with PS as a new beam source will be reduced to 1/4 - 1/5 of a traditional FEL.

The pseudospark discharge has many potential applications, it can be used for the free electron lasers, the X-ray pumping, the high power switch, the electron beam lithography and the plasma processing.

## REFERENCES

1. C. Cheng, Z.J. Wang, et al, The Raman FEL Investigation, Chinese Science, A29, 992(1986).

2. B. Feng, M.C. Wang, et al, A Novel Small-period Wiggler and Simulation of its FEL, IEEE J. Quant. Electr. 29(8), 2263(1993).

3. M.C. Wang, Z.J. Wang, et al, Experiments of a Raman FEL with Distributed Feedback Cavity, Nucl. Instr. Mech. A304, 116(1991).

4. J. Christiansen and Ch. Schutheiss, Production of High-current Particle Beams by Low Pressure spark Discharge, Z. Phys., A290, 35(1979).

5. M.C. Wang, J.B. Zhu, Z.J. Wang, Design of Pseudospark Discharge for Raman FEL, Nucl. Instr. Mech. A358, 38(1995).

6. M.J. Rhee and B.N. Ding, Time-resolved Energy Spectrum of Pseudospark-produced Electron Beam, Phys. Fluids B4(30).764(1992)

7. W. Bauer, et al, Review on Effect of the High Power Pseudospark, in Proc. Seventh Inter. Conf. on High-power Particle Beams, 233(1988).

8. M.C. Wang, J.B. Zhu, Z.J. Wang, et al, Successful Operation of Pseudospark Electron Beam Source, Acta Optica Sinica, 15(2),255(1995).

9. M.C. Wang, Z.J. Wang, Proposal of a Compact FEL, to be published in Acta Optica Sinica, (1995).

Table 1. Parameter Values of the Raman FEL at SIOFM

| Accelerator | | |
|---|---|---|
| Beam voltage | V | 400 KV |
| Beam current | I | 400 A |
| Beam radius | $r_c$ | 3 mm |
| Pulse duration | $\tau$ | 60 ns |
| Wiggler | | |
| Wiggler period | $\lambda_w$ | 10 mm |
| Wiggler length | L | 600 mm |
| Wiggler field | $B_w$ | 1500 G |
| Internal radius | $r_i$ | 8 mm |
| External radius | $r_0$ | 14 mm |
| Copper-sheet thickness | $h_1$ | 1.5 mm |
| Iron-cores thickness | $h_2$ | 3.5 mm |
| Radiation wave | | |
| Wavelength | $\lambda_s$ | 3 mm |
| Output power | $P_0$ | 1 MW |
| Pulse duration | | 15 ns |
| Efficiency | $\eta$ | 0.63 % |

Figure Captions

Fig.1 Configuration of the helical small period wiggler with
ferro-magnetic cores. (a), Schematic of the wiggler;
(b), Cross section of the wiggler along z-axis.

Fig.2 Grating spectrometer data showing the spectrum of superradiant
amplifier(dashed line) and oscillator(solid line) for
$B_0 = 9.45$ KG, $B_w = 1.26$ KG with 0.4MeV/ 800 A beam.

Fig.3 A typical Paschen curve with numerical simulations.

Fig.4 The distance d (cm) is a function of the pressure p at
a given voltage.

Fig.5 The experimental configuration of the pseudospark discharge.

Fig.6 The PS voltage waveform, peak voltage 200 KV(84 KV/mm),
pulse duration $< 3$ $\mu s$(0.3 $\mu s$/cycle).

Fig.7 The optical density profile.

Fig.8 The schematic of the Raman FEL.

Submit to the Symposium/Workshop on Development and Application of
FEL, CCAST, May 28 - June 3(1995)

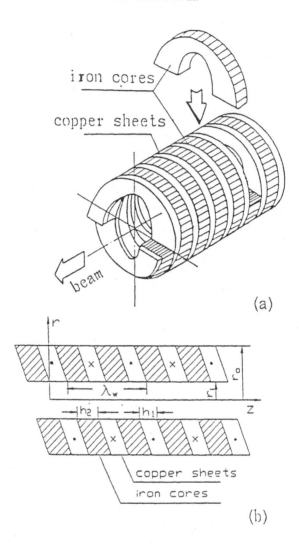

(a)

(b)

# Fig. 1

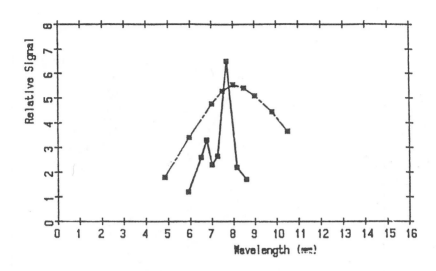

## Fig. 2

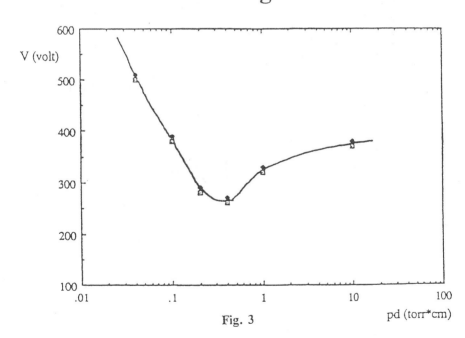

Fig. 3

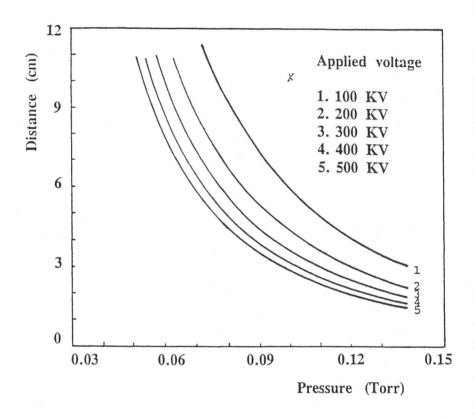

**Fig. 4**

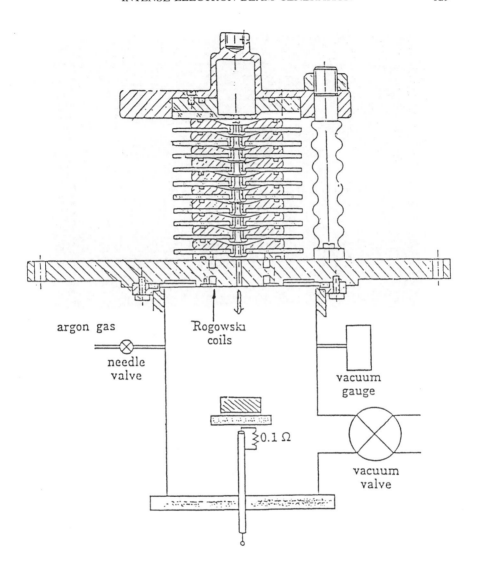

# Fig. 5

M.C. WANG *et al.*

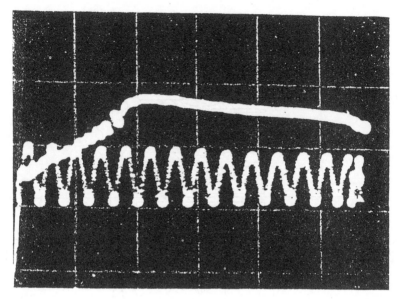

Fig. 6

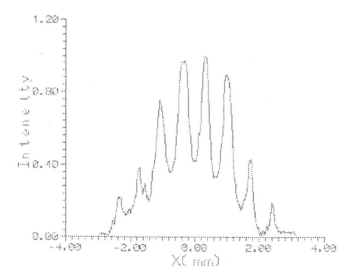

Fig. 7

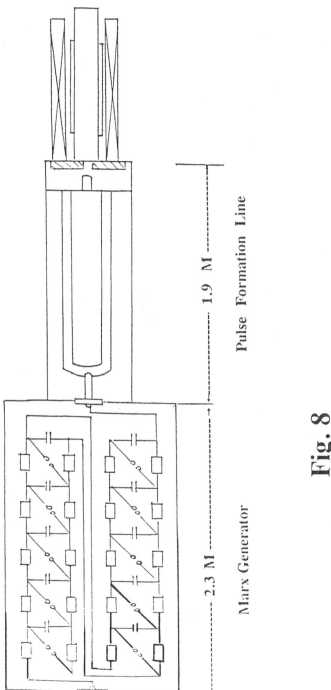

Fig. 8

# Free Electron Lasers

Claudio Pellegrini

Physics Department, University of California at Los Angeles
Los Angeles, California, 90095

March 5, 1996

### Abstract

We review the physics of Free Electron Lasers, and the collective FEL instability that characterizes this system. We show that the instability can be described with a single parameter, the FEL parameter, that determines the instability growth rate (or FEL gain), and the saturation length and power. We also discuss the physics and technology issues relevant for the operation of an FEL at a wavelength of about 0.1nm, and the expected peak power, average power, and brightness of such a source. We show that based on our present understanding of FEL physics, and using existing technology, we can build a 0.1nm FEL with peak power of about 10 GWatt, pulse duration of about 0.1 ps, and bandwidth of about 0.1%.

# 1 INTRODUCTION

Free Electron Lasers (FELs) are powerful sources of coherent electromagnetic radiation, operating in the microwave to UV region of the spectrum. Recent progress in the FEL physics and technology makes now possible to extend their wavelength into the Soft X-ray and X-ray region, with peak power and brightness many orders of magnitude larger than that obtainable from other sources.

In this paper we review the basic properties of FELs, starting from the FEL collective instability [1], and relate the FEL growth rate to the electron beam phase-space density. We then discuss the development of FELs as a high power, coherent source of Soft X-ray and X-ray radiation. This development is made possible by the recent progress in the production of high brightness electron beams, and by the use of the Self Amplified Spontaneous Emission (SASE) mode of operation of the FEL [1], [2], [3].

In the SASE mode lasing is produced in a single pass of a high phase-space density electron beam through a long undulator, eliminating the need for optical cavities, difficult to build in the Soft X-ray or X-ray spectral region. However in the SASE approach the requirements on the electron beam peak current, emittance, and energy spread are very stringent [5], [4] and until recently difficult to satisfy. This situation has been changed by the recent development of high-brightness radio frequency photocathode electron guns [6], and the progress in accelerating and compressing these beams without spoiling their brightness resulting from the work on linear colliders [7],[8], [9],[10], [11]. As a result there is now the possibility to make a major extension of FEL operation, from the shortest wavelength yet achieved -240 nm- to 0.1 nm [12], [13], [14], [15], [16], [17].

In this paper we will first review the spontaneous emission of radiation from an electron crossing an undulator magnet, the basic element of FEL physics. We will then discuss the production of coherent undulator radiation, and its connection to FELs. The next section will be a review of the 1-dimensional model of the FEL, with a discussion of the FEL collective instability, the derivation of the exponential gain, and of the effect of the beam energy spread on the gain.

# 2   Principle of operation

## 2.1   Undulator radiation from one electron

The FEL is based on the emission of radiation from relativistic electrons moving in an undulator magnetic field (undulator radiation). We will first review the basic characteristics of the undulator radiation from a single electron; we will then discuss how the undulator radiation intensity is increased in an FEL, using high brightness electron beams. We assume for simplicity the undulator to have a helical magnetic field, of amplitude $B_u$, and period $\lambda_u$. A more detailed discussion, including the case of planar undulators, can be found in reference [18], to which we refer the reader. If $z$ is the undulator axis, and $x, y$, the two perpendicular directions, the undulator field, near to the axis, is approximately

$$B_x = B_u \cos(2\pi \frac{z}{\lambda_u}), \ B_y = B_u \sin(2\pi \frac{z}{\lambda_u}). \tag{1}$$

One relativistic electron of energy $E = mc^2\gamma$, and momentum $p_z >> p_x, p_y$, traversing the undulator, executes a helical trajectory, with constant velocity $V_z$, and transverse velocity

$$\frac{V_x}{c} = \frac{a_u}{\gamma} \sin(2\pi \frac{z}{\lambda_u}), \frac{V_y}{c} = \frac{a_u}{\gamma} \cos(2\pi \frac{z}{\lambda_u}), \tag{2}$$

where

$$a_u = \frac{eB_u\lambda_u}{2\pi mc^2}, \tag{3}$$

is called the undulator parameter, and is the undulator vector potential normalized to the electron rest energy. The helix radius is

$$R = 2\pi \frac{a_u\lambda_u}{\gamma}. \tag{4}$$

For a relativistic beam the periodic magnetic field of the undulator appears approximately, using the Weiszacker-Williams approximation, as a circularly polarized plane wave, of wavelength $\lambda = \lambda_u/\gamma_z$, where $\gamma_z = \gamma/\sqrt{1 + a_u^2}$ is the relativistic factor for the Lorentz transformation to the frame where the longitudinal momentum of the electron is zero. Some of the photons in this plane wave can be backscattered by the electrons, and in the laboratory frame they will appear as spontaneous radiation, emitted in a narrow line centered at the wavelength

$$\lambda = \frac{\lambda_u}{2\gamma^2}(1 + a_u^2). \tag{5}$$

The additional factor $\gamma_z$ in (5) appears when transforming again to the Laboratory system

The spontaneous radiation is emitted in a narrow cone of aperture $1/\gamma$ around the axis. The width of the radiation line (bandwidth) is related to the number of undulator periods $N_u$ [18] by

$$\frac{\Delta\omega}{\omega} = \frac{1}{2N_u}. \tag{6}$$

The undulator is an extended linear source, but it can be approximately described as an equivalent source at the undulator center, with angular aperture

$$\theta = \sqrt{\frac{\lambda}{\lambda_u N_u}}, \tag{7}$$

and an effective source radius ( diffraction limited)

$$a = \frac{1}{4\pi}\sqrt{\lambda\lambda_u N_u}. \tag{8}$$

Notice that the product

$$a\theta = \frac{\lambda}{4\pi}, \tag{9}$$

gives the minimum phase space for a diffraction limited photon beam.

The intensity of the radiation emitted on axis, and at the wavelength (5) is [18]

$$\frac{d^2 I}{d\omega d\Omega} = 2N_u^2 \frac{e^2}{c}\gamma^2 \frac{a_u^2}{(1 + a_u^2)^2}. \tag{10}$$

The coherent intensity is obtained by multiplying (10) by the solid angle corresponding to (7) and the bandwidth (6). Dividing this intensity by the energy of a photon with the wavelength (5) we can also rewrite the coherent intensity as the number of photons per electron within the solid angle corresponding to (7) and bandwidth (6) as

$$N_{ph} = \pi\alpha\frac{a_u^2}{1 + a_u^2}, \tag{11}$$

where $\alpha$ is the fine structure constant. For a typical value $a_u \approx 1$ one obtains $N_{ph} \approx 10^{-2}$, showing that the undulator radiation process is rather inefficient.

The number of photons per electron is the basic number determining the brightness of an undulator source. In fact assuming that the electron beam transverse radius and angular divergence are smaller than that of the effective radiation source defined previously, the brightness is given by

$$B_\lambda = 4\pi^2 \frac{N_e \, N_{ph}}{\lambda^2}, \tag{12}$$

where $N_e$ is the number of electrons per second, proportional to the average electron current. In the opposite case, which can be characterized by the condition that the beam emittance $\epsilon$, the product of the electron beam transverse radius and angular divergence, is larger than the corresponding quantity for the radiation beam, $\lambda/4\pi$, the brightness becomes

$$B_\lambda = \frac{N_e \, N_{ph}}{\epsilon^2}. \tag{13}$$

The conventional definition of brightness considers also the frequency spread of the radiation. assuming a 0.1% spread as the reference. The previous definition then becomes

$$B_\lambda = \frac{N_e \, N_{ph}}{\epsilon^2} \frac{10^{-3}}{\Delta\omega/\omega}. \tag{14}$$

The transverse electron beam brightness is defined in a similar way as

$$B_{e,T} = e\frac{N_e}{\epsilon^2}. \tag{15}$$

In defining the brightness it is important to distinguish between the peak brightness and the average brightness, depending on wether we use the peak or average value of $N_e$. For instance the peak electron beam current is what determines the FEL gain and performance. Similarly the peak photon brightness is important for X-ray imaging, while the average brightness might be important for other applications.

## 2.2   FEL radiation

To increase the peak brightness for a given wavelength we can either increase the electron current or increase the number of photons produced per electrons. An FEL achieves the second goal, by increasing the number of photons per electron by about seven orders of magnitude.

How do we increase the number of photons emitted per electron? If we have many electrons, say $N_e$, and they are all grouped within a small fraction of a wavelength, the total intensity would be the single particle intensity times $N_e$, and the number of photons per electron would be increased by a factor $N_e$. In practical cases the electrons are in a bunch much longer than the radiation wavelength, and their position distribution on a scale of $\lambda$ is completely random. As a result the radiation fields emitted by different electrons have a random relative phase, the total intensity is proportional to $N_e$, and the number of photons per electrons is still given by (12).

However also in this case we can increase the number of photons emitted per electron if we take advantage of a collective instability of the electron beam-EM radiation field-undulator system [1]. This instability works as follows:

1. the electron beam interacts with the electric field of the radiation; the electric field is perpendicular to the direction of propagation of the beam (the undulator axis), and is parallel to the wiggling (transverse) velocity (2) of the electrons produced by the undulator magnet, of amplitude $a_u/\gamma$; the interaction produces an electron energy modulation, on the scale $\lambda$;

2. the electron energy modulation modifies the electron trajectory in the undulator, in a such a way to produce bunching of the electrons at the scale $\lambda$;

3. electrons bunched within a wavelength emit radiation in phase, thus producing a larger intensity; the larger intensity leads to more energy modulation and more bunching, leading to exponential growth of the radiation; the intensity can reach the limit $I \sim N_e^2$ for the case of extreme bunching (superradiance) [19]

The FEL instability will be discussed in detail in the following section, in the simple case of a one dimensional theory. Here we will summarize some of the most important results for the FEL physics. For the collective instability to occur there are some conditions that must be satisfied . These conditions depend on the FEL parameter [1],

$$\rho = (\frac{a_u}{4\gamma}\frac{\Omega_p}{\omega_u})^{2/3}, \tag{16}$$

and on the beam emittance, $\epsilon$ [4], [5]. The quantities in (16) are: $\omega_u = 2\pi c/\lambda_u$ is the frequency associated to the undulator periodicity,

$$\Omega_p = (\frac{4\pi r_e c^2 n_e}{\gamma})^{1/2}, \tag{17}$$

is the beam plasma frequency, $n_e$ is the electron density, and $r_e$ is the classical electron radius. The FEL parameter characterizes the instability, giving the instability growth rate, or gain length,

$$L_G \approx \frac{\lambda_u}{2\sqrt{3}\pi\rho} \tag{18}$$

The amplitude of the radiation field grows exponentially along the undulator axis z, as $A \approx A_0 \exp(z/L_G)$. The conditions are:

a. beam emittance smaller than the wavelength:

$$\epsilon < \frac{\lambda}{4\pi} \tag{19}$$

b. beam energy spread smaller than the FEL parameter:

$$\sigma_E < \rho \tag{20}$$

c. undulator length larger than the gain length:

$$N_u \lambda_u >> L_G \tag{21}$$

d. the gain length must be shorter than the radiation Raleigh range:

$$L_G < L_R, \tag{22}$$

where the Raleigh range is defined in terms of the radiation beam radius, $w_0$, and the wavelength by $\pi w_0^2 = \lambda L_R$.

Condition $a$ says that for the instability to occur the electron beam must match the angular and transverse space characteristics of the radiation emitted by one electron in traversing the undulator, equations (7, 8, 9). This is also the condition to obtain from the electron beam diffraction limited spontaneous radiation. Notice that for nanometer wavelength this condition cannot be met at present by storage ring based synchrotron radiation sources, but it can be satisfied by electron beams produced by a Radio Frequency laser driven electron guns (photoinjectors), as we will discuss in the following sections.

Condition $b$ limits the beam energy spread to a value such that the width of the spontaneous radiation line is not increased. Conditions $c$ introduces a requirement on the minimum undulator length for this process to become significant. Condition $d$ requires that more radiation is produced by the beam than what is lost through diffraction out of the finite radius beam. Conditions $a$ and $d$ both depend on the beam radius and the radiation wavelength , and are not independent. If they are satisfied diffraction and 3-dimensional effects are not important, and we can use with good approximation the 1-dimensional model.

If these conditions are satisfied the radiation field emitted by the beam will grow exponentially along the undulator length, with a growth rate given by (18). The field will saturate after an undulator length ( saturation length, ) of the order of ten gain length. At saturation the radiation power is given by [1]

$$P_{sat} = \rho I_{beam} E_{beam}, \tag{23}$$

where $I_{beam}$ is the beam current, and $E_{beam}$ the beam energy, and the number of photons per electron is

$$N_{sat} = \rho \frac{E}{E_{ph}} \tag{24}$$

If we consider a case of interest to us for a Soft X-ray FEL, with $E_{ph} \approx 250eV$, $E \approx 3GeV$, $\rho \approx 10^{-3}$, we obtain $N_{ph} \approx 10^4$, i.e. an increase of almost 6 orders of magnitude in the number of photons produced per electron. This increase is reflected in a much larger brightness.

# 3   The 1-dimensional FEL theory

AS we discussed in the previous section the system consisting of an electron beam, an electromagnetic field and an undulator, is an unstable system if the beam has a longitudinally uniform distribution. The system will evolve toward a state in which the electrons are bunched at the radiation wavelength, i.e. they are equally spaced in the longitudinal direction and separated by a distance equal to $\lambda$. To see how this transition from the initial to the final state can take place we have to study the Maxwell equations for the electromagnetic field, in combination with the equation of motion for the electrons in the combined field of the undulator magnet and of the radiation field. we will follow closely the work of references [1], [18]. Let us look at the equation describing the electron energy change in an E-M field of amplitude $A$ and phase $\Psi$,

$$mc^2 \frac{d\gamma}{dt} = eA\frac{a_u}{\gamma} \sin(\Phi + \Psi) \tag{25}$$

where the phase is

$$\Phi = 2\pi(\frac{1}{\lambda} + \frac{1}{\lambda_u})\beta_z - \omega. \tag{26}$$

If $z = \beta_z ct + z_0$, and the distribution of $z_0$ is uniform, as is usually the case for an electron beam produced in an accelerator, then the phase covers all values between 0 and $2\pi$, and some particles will gain energy, while some will loose energy. The change in energy will result in a change of velocity, which will produce a change in phase. Hence the system will evolve.

An equilibrium state for the system exists, and corresponds to electrons being at the phase $n\pi$, which gives zero energy change, and also which corresponds to all electrons having the resonant energy. Hence if we start from the initial state with a uniform distribution in phase and in longitudinal electron distribution, the system will evolve toward the "bunched" equilibrium state. The speed at which the system will evolve will define the gain length.

To describe this situation and this evolution we need to introduce the dynamical variables describing the electron-electromagnetic field system. For simplicity we consider only the case of a helical undulator.

We also make the following approximations:

1. we assume that the beam transverse size is much larger than the radiation wavelength, and use a 1-dimensional picture, neglecting diffraction; the E-M field is described as a plane wave and the beam transverse density distribution is assumed constant;

2. We assume that the field is propagating in vacuum;

3. we assume that the E-M field, oscillating at the frequency $\omega = 2\pi c/\lambda$, has an amplitude and a phase that change slowly, and simplify Maxwell equations neglecting their second derivatives;

4. we neglect the slippage, $S = N_u\lambda$, between electrons and the E-M field due to the difference in velocity, assuming that it is smaller than the bunch length, $L_e$, or $S/L_e \ll 1$;

5. We neglect the effects due to the beam emittance.

The E-M Field is described with the vector potential of the undulator and of the radiation field,

$$\vec{A} = \vec{A}_u + \vec{A}_R . \tag{27}$$

Using circularly polarized waves and a helical undulator we have

$$\vec{A}_u = -\frac{B_u}{k_u}(\cos(k_u z)\ \vec{x} + \sin(k_u z)\ \vec{y}), \tag{28}$$

where $k_u = 2\pi/\lambda$, and

$$\vec{A}_R = -\frac{A(z,t)}{k_R}(-\cos(\phi_R)\ \vec{x} + \sin(\phi_R)\ \vec{y}), \tag{29}$$

where $k_R = 2\pi/\lambda$ ,

$$\phi_R = k_R(z - ct) + \psi(z,t), \tag{30}$$

and $A(z,t)$, and $\psi(z,t)$ are the slowly varying amplitude and phase of the wave. The system Hamiltonian is

$$H = c[(\vec{P} - \frac{e}{c}\ \vec{A})^2 + m^2c^2]^{1/2}, \tag{31}$$

and does not depend on $x$ and $y$. Hence the corresponding canonical momenta are constants of the motion. From these constants we can obtain the electron transverse velocity

$$\beta_x = \frac{e}{mc^2\gamma}(A_{u,x} + A_{R,x}) + \frac{P_{x,0}}{mc\gamma}, \tag{32}$$

and

$$\beta_y = \frac{e}{mc^2\gamma}(A_{u,y} + A_{R,y}) + \frac{P_{y,0}}{mc\gamma}. \tag{33}$$

Introducing the quantities

$$a_u = \frac{eB_u}{mc^2k_u}, \tag{34}$$

$$a_R = \frac{eA(z,t)}{mc^2 k_R}, \tag{35}$$

we also have

$$\beta_x = \frac{1}{\gamma}[a_u \cos(k_u z) - a_R \cos(\phi_R)] + \frac{P_{x,0}}{mc\gamma}, \tag{36}$$

$$\beta_x = \frac{1}{\gamma}[a_u \sin(k_u z) + a_R \sin(\phi_R)] + \frac{P_{y,0}}{mc\gamma} \tag{37}$$

Notice that for typical cases $a_u \approx 1, a_R \ll 1$.

In the approximation of neglecting emittance effects, we will assume for the time being that the two initial values of the canonical momenta, $P_{x,0}, P_{y,0}$ can be neglected.

We could now use the Hamiltonian to obtain the equations of motion for $z$, and $P_z$, but it is simpler to use instead the equivalent equations for the energy change, and for the phase change

$$mc^2 \frac{d\gamma}{dt} = \frac{eA(z,t)a_u}{\gamma} \sin(\Phi + \psi(z,t)), \tag{38}$$

$$\frac{d\Phi}{dt} = (k_R + k_u)z - \omega, \tag{39}$$

and obtain the longitudinal velocity from the energy, using the relationship

$$\beta_z = (1 - \frac{1}{\gamma^2} - \beta_x^2 - \beta_y^2)^{1/2}. \tag{40}$$

For relativistic velocities this can be approximated as

$$\beta_z \approx 1 - \frac{1}{2\gamma^2} - \frac{\beta_x^2 + \beta_y^2}{2} = 1 - \frac{1}{2\gamma^2}(a_u^2 + a_R^2 - 2a_u a_R \cos(\Phi + \psi)). \tag{41}$$

Next we need the equations for the field. We write Maxwell equations in the form

$$(\nabla^2 - \frac{1}{c^2}\frac{\partial^2}{\partial t^2})\vec{A} = -\frac{4\pi}{c}\vec{J}, \tag{42}$$

$$(\nabla^2 - \frac{1}{c^2}\frac{\partial^2}{\partial t^2})V = -4\pi\rho. \tag{43}$$

The current and charge density are

$$\rho = e \sum \delta(\vec{r} - \vec{r_i(t)}), \tag{44}$$

$$\vec{J} = ec \sum \vec{\beta_i}\, \delta(\vec{r} - \vec{r_i(t)}). \tag{45}$$

We consider first the effects of the transverse fields, and introduce the quantities

$$\hat{J} = J_x - iJ_y = ec \sum \delta(\vec{r} - r_i(t))\frac{1}{\gamma_i}[a_u e^{-ik_u z} - a_R e^{i\phi_R}], \tag{46}$$

$$\overset{\wedge}{A} = A_x - iA_y = i\frac{\alpha}{k_R}e^{ik_R(z-ct)}, \text{ with } \alpha(z,t) = -iA(z,t)e^{i\psi}. \qquad (47)$$

In our approximation of a plane wave the sum over particles can be rewritten as

$$\sum \delta(\vec{r} - r_i(t)) \approx \sigma_0 e \sum \delta(z - z_i(t)), \qquad (48)$$

where $\sigma_0$ is the transverse beam density.

With these definitions, and using the slowly varying amplitude and phase approximation, the Maxwell equations for the complex amplitude $\alpha$ become

$$(\frac{\partial}{\partial z} - \frac{1}{c}\frac{\partial}{\partial t})\alpha = 2\pi\sigma_0 \sum_{l=1}^{N} \frac{1}{\gamma_l} e \sum \delta(z - z_l(t))(a_u e^{-i\Phi_l} - \frac{ie\alpha}{mc^2 k_R}). \qquad (49)$$

We write now the complete set of 1-D equations, using complex notations, and introducing the "resonant longitudinal velocity" $\beta_{zR}$ and the resonant energy $\gamma_R$ such that

$$\lambda = \lambda_u \frac{1 - \beta_{zR}}{\beta_{zR}} = \lambda_u \frac{1 + a_u^2}{2\gamma_R^2}. \qquad (50)$$

.Notice that to establish the relationship between the resonant longitudinal velocity and the resonant energy we assume that the electrons are ultrarelativistic, and that the transverse velocity is determined only by the undulator magnetic field and not by the radiation field. The set of equations is

$$\frac{d}{dt}\Phi_l = k_u(1 - \frac{\gamma_R^2}{\gamma^2}) + \frac{k_R}{2\gamma^2}\{a_u(i\alpha e^{i\Phi_l} - cc) - |\alpha|^2\}, \qquad (51)$$

$$\frac{d}{dt}\gamma_l = \frac{a_u k_R}{2\gamma_l}(\alpha e^{i\Phi_l} + cc), \qquad (52)$$

$$(\frac{\partial}{\partial z} + \frac{1}{c}\frac{\partial}{\partial t})\alpha = 2\pi\sigma_0 \sum_{l=1}^{N} \frac{1}{\gamma_l}\delta(z - z_l)(a_u e^{-i\Phi_l} - \frac{ie\alpha}{mc^2 k_R}). \qquad (53)$$

## 3.1   The small slippage regime

The field equations can be simplified in the small slippage regime. Due to the difference in velocity between the electrons and the radiation field, the field will advance in front of the electrons by one wavelength per undulator period. We call slippage the quantity

$$S = N_u\lambda, \qquad (54)$$

$N_u$ being the number of periods in the undulator. To characterize different FEL regimes we use the ratio of the slippage to the electron bunch length $L_e$.

The FEL regimes can be characterized according to wether this ratio is larger or smaller than one. In this paper we will consider only the case in which the ratio is much less than one. For a discussion of the general case the reader is referred to reference [19].

To consider the effect of the slippage we write the electromagnetic field amplitude as a function of $t$ and $z - V_g t$ , where $V_g$ is the group velocity

$$\alpha = \alpha(t, \frac{z - V_G t}{L}). \tag{55}$$

In what follows we will assume to simplify our discussion that $V_g = c$. The dependence on $z - ct$ is used to describe the shape of the wave packet, which we assume to have a length L, on the order of the electron bunch length, $L_e$ , and the field amplitude can be written as

$$\alpha(t, z - ct) = \sum \alpha_n(t) e^{i2\pi n(z - ct)/L}. \tag{56}$$

If we evaluate the argument $z - ct$, at the particle position $z = z_l(t) = Vt + z_{l0}$ we have, using (5.54),

$$\frac{z_l - ct}{L} = -\frac{ct(1 - \beta_z)}{L} + \frac{z_{l0}}{L} < \frac{N\lambda_u}{2L\gamma^2} + \frac{z_{l0}}{L} \approx \frac{S}{L} + \text{constant.} \tag{57}$$

If we are in the small slippage case, $S/L \ll 1$, the argument $z - ct$ in (55) is a constant, and we can approximate $\alpha(t, z - ct) \approx \alpha(t, \text{constant})$ and the only important component of the expansion in (56) is $\alpha_0(t)$.

The field equation now becomes

$$\frac{\partial}{\partial t}\alpha_0(t) = 2\pi \frac{ec\sigma_0}{L}\{a_u N_e B(t) - \frac{ie\alpha_0}{mc^2 k_R}\sum_{l=1}^{N_e}\frac{1}{\gamma_l}\}, \tag{58}$$

where

$$B(t) = \frac{1}{N}\sum_{l=1}^{N_e} e^{-i\Phi_l} \tag{59}$$

is the bunching factor, which plays a key role in the FEL physics.

## 3.2   The normalized FEL equation

We now put the FEL equations in a form that facilitate their analysis, using the notations introduced in ([1]).

We introduce the beam plasma frequency

$$\Omega_p = (\frac{4\pi r_e c^2 n_e}{\gamma})^{1/2}, \tag{60}$$

the FEL parameter

$$\rho = (\frac{a_u}{4\gamma}\frac{\Omega_p}{\omega_u})^{2/3}, \tag{61}$$

and the detuning parameter

$$\delta = \frac{\gamma^2 - \gamma_R^2}{2\rho\gamma_R^2}, \tag{62}$$

where $\gamma$, $\gamma_R$ are the beam energy and the resonant energy. Notice that using the relationship between energy and wavelength of the emitted radiation, equations (5, 52), the quantity $\delta$ can also be written as

$$\delta = \frac{\lambda - \lambda_R}{4\lambda_R\rho}, \tag{63}$$

and represents the detuning of the FEL measured in units of the FEL parameter $\rho$.

We then introduce the scaled variables

$$\tau = 4\pi\rho\frac{\gamma_R^2}{\gamma^2}\frac{ct}{\lambda_u}, \tag{64}$$

$$\Gamma_l = \frac{\gamma_l}{\rho\gamma}, \tag{65}$$

$$\Psi_l = \Phi_l - \omega_u(1 - \frac{\gamma_R^2}{\gamma^2})t, \tag{66}$$

$$\overset{\wedge}{A} = \frac{ea_u}{4mc^2\gamma_R^2 k_u\rho^2}\alpha_0 e^{i\omega_u(1-\frac{\gamma_R^2}{\gamma^2})t}. \tag{67}$$

With these notations the FEL equations are:
the phase equation

$$\frac{d\Psi_l}{d\tau} = \frac{1}{2\rho}(1 - \frac{1}{\rho^2\Gamma_l^2}) + \frac{i}{\rho\Gamma_l^2}(\overset{\wedge}{A} e^{i\Psi_l} - cc) - 2\rho\frac{(1 + a_u^2)}{a_u^2\Gamma_l^2}|\overset{\wedge}{A}|^2, \tag{68}$$

the energy exchange equation

$$\frac{d\Gamma_l}{d\tau} = -\frac{1}{\rho\Gamma_l}\{\overset{\wedge}{A} e^{i\Psi_l} + cc\}, \tag{69}$$

the field equation

$$\frac{d\overset{\wedge}{A}}{d\tau} = <\frac{e^{-i\Psi}}{\Gamma}> + i\delta \overset{\wedge}{A} - 2i \overset{\wedge}{A} \rho\frac{1 + a_u^2}{a_u^2} < \frac{1}{\Gamma} >, \tag{70}$$

where $<> = (1/N_e)\sum_{l=1}^{N_e}$.

In most cases the FEL parameter is small and we can simplify the equations as

$$\frac{d\Psi_l}{d\tau} = \frac{1}{2\rho}(1 - \frac{1}{\rho^2\Gamma_l^2}), \tag{71}$$

$$\frac{d\Gamma_l}{d\tau} = -\frac{1}{\rho\Gamma_l}(\overset{\wedge}{A}\, e^{i\Psi_l} + cc),$$ (72)

$$\frac{dA}{d\tau} = <\frac{e^{-i\Psi}}{\Gamma}> + i\,\overset{\wedge}{A}\,\delta,$$ (73)

which we will use as the basic FEL equations.

## 3.3 Hamiltonian and conservation laws.

The FEL equations have two conserved quantities:

$$E = |\overset{\wedge}{A}|^2 + <\Gamma>,$$ (74)

and

$$H = \sum_l \{\frac{1}{2\rho^2}(\rho\Gamma_l + \frac{1}{\rho\Gamma_l}) - \frac{i}{\rho^2\Gamma_l}(\overset{\wedge}{A}\, e^{i\Psi_l} + cc) - |\overset{\wedge}{A}|^2(\delta - 2\rho\frac{1+a_u^2}{a_u^2}\frac{1}{\Gamma_l})\}.$$ (75)

The first quantity is the total energy, which can also be written as

$$E = mc^2 n_e\gamma + \frac{1}{4\pi}|A|^2 = constant,$$ (76)

telling us that the beam kinetic energy density plus the field energy density is a constant;

The FEL equations of motion can be obtained from the Hamiltonian in the standard form, considering $\Gamma, \Psi, \overset{\wedge}{A}, \overset{\wedge}{A}^*$ as conjugate variables.

## 3.4 The FEL collective instability

To determine the behavior of the FEL when the initial state is an electron beam with a uniform longitudinal phase distribution, a given energy spread, and zero initial field amplitude, we linearize the FEL equations, assuming

$$\Gamma = 1 + \eta, \eta << 1,$$ (77)

and neglecting terms in the square of the field, to obtain

$$\frac{d\Psi_l}{d\tau} = \frac{\eta_l}{\rho} + i\rho(\overset{\wedge}{A}\, e^{i\Psi_l} - cc),$$ (78)

$$\frac{d\eta_l}{d\tau} = -\rho(1 - \eta_l)(\overset{\wedge}{A}\, e^{i\Psi_l} + cc),$$ (79)

$$\frac{d\overset{\wedge}{A}}{d\tau} = i\delta\,\overset{\wedge}{A} + <(1 - \eta_l)e^{i\Psi_l}>.$$ (80)

This equation can be obtained from the Hamiltonian

$$H = \sum_l \{ \frac{\eta_l^2}{2\rho} - i\rho(1-\eta_l)(\overset{\wedge}{A}\, e^{i\Psi_l} - cc) \} - \delta\, \overset{\wedge}{A}\overset{\wedge}{A}^{*} \tag{81}$$

We characterize the beam by a distribution function $f = f(\Psi, \eta, \tau)$ , satisfying Vlasov equation

$$(\frac{\partial}{\partial\tau} + \frac{d\Psi_l}{d\tau}\frac{\partial}{\partial\Psi} + \frac{d\eta}{d\tau}\frac{\partial}{\partial\eta})f = 0. \tag{82}$$

We assume the distribution function to be

$$f = \frac{1}{2\pi} f_0(\eta) + f_1(\eta)e^{i\Psi + i\mu\tau}, \tag{83}$$

and use it to evaluate the current term in the field equation. We also assume the field to be proportional to $e^{i\mu\tau}$. We obtain

$$\overset{\wedge}{A} = -\frac{2\pi i e}{\mu - \delta} e^{i\mu\tau} \int f_1(\eta)(1-\eta)d\eta. \tag{84}$$

Notice that $\overset{\wedge}{A} = 0, f = f_0$ is a solution of the Vlasov equation. Substituting the expression for the field in the Vlasov equation we obtain the dispersion relation

$$\mu - \delta + \rho \int d\eta \frac{(1-\eta)^2}{\mu + (\eta/\rho)}\frac{\partial f_0}{\partial\eta} = 0. \tag{85}$$

For a monochromatic beam, the dispersion relation becomes

$$\mu^3 - \delta\mu^2 + 2\rho\mu + 1 = 0. \tag{86}$$

The system is unstable when the dispersion relation has solutions with $\text{Im}\,\mu < 0$. For a monochromatic beam such solutions exist if $\delta < 1.93$. The $\text{Im}\,\mu$ has a maximum when $\delta = 0$. If $\rho$ is small this maximum is given by the root of $\mu^3 + 1 = 0$, and is $\sqrt{3}/2$.

To see the effect of a spread in energy of the electron beam we have evaluated the dispersion relation for the case of a Lorentzian distribution

$$f_0(\eta) = \frac{1}{\pi}\frac{\Delta}{\eta^2 + \Delta^2} \tag{87}$$

The results are shown in figure 1, where we show the imaginary part of the eigenvalue, the growth rate, for three different values of the ratio of the energy spread, $\Delta$, to $\rho$.

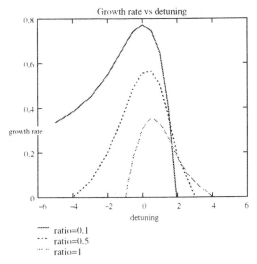

Figure 1.

The figure shows that the growth rate decreases when the energy spread becomes of the order of the FEL parameter $\rho$. One can also notice that for a small energy spread there is threshold for the instability at $\delta \approx 2$. This threshold disappears for larger values of the energy spread. When the energy spread becomes larger than $\rho$, the growth rate becomes proportional to the derivative of the unperturbed distribution function, $\mathrm{Im}\,\mu \approx \partial f_0/\partial \eta$ [18]. The dependence of the growth rate on the slope of the distribution function can be considered analogous to the population inversion in quantum lasers.

One can see from this picture that for an energy spread small compared to $\rho$, the growth rate, in terms of the normalized time $\tau$, is of the order of one. Going back to the laboratory time, and using (64) with the usual case $\gamma \approx \gamma_R$, we have an exponential growth rate that for $\rho$ small, and $\delta = 0$ is

$$L_G = (ct)_G = \frac{\lambda_u}{2\sqrt{3}\pi\rho} \tag{88}$$

in agreement with (18).

The exponential growth described in the linear approximation, must saturate at some field level. For small values of the FEL parameter $\rho$, and assuming the field to be nearly constant, as is the case at saturation, the equations (81,82), describing the electron motion in the energy-phase plane, are like a pendulum equation. The FEL saturates when the growth rate becomes of the order of the period of rotation of the electrons in the energy-phase plane. Evaluating this condition we obtain that saturation occurs after about ten gain lengths.

At saturation the normalized field amplitude $\hat{A}$ is also of the order of one, and the beam energy spread is of the order of $\rho$ [1]. From the result $\hat{A} \approx 1$, and using (74) we also obtain eq. (23) for the saturation power.

## 3.5  Small signal gain

While for long undulators, $N_u \lambda_u > L_G$, only the eigenvalue of the dispersion relation with negative imaginary part is important, for short undulators, $N_u \lambda_u < L_G$, all three eigenvalues of the dispersion relation (87) are important in determining the field amplitude, and in fact the interference between these three waves determines the change in the field amplitude. In this case, which is called the small signal regime case, the gain, defined as the change in field intensity, $I$, over the intensity for one undulator crossing,

$$G_{ssg} = \frac{\Delta I}{I}, \tag{89}$$

is given by ([20])

$$G_{ssg} = \frac{4}{\delta^3}(1 - \cos(\delta\tau) - \frac{\delta\tau}{2}\sin(\delta\tau)) \tag{90}$$

where, using (65, 66)

$$\delta\tau = 4\pi N_u \frac{\Delta\omega}{\omega} \tag{91}$$

The small signal gain vs $\delta$ is shown in figure 2 for the case $\tau = 1$.

Notice that this curve is antisymmetric respect to $\delta$, while the similar curve for the gain length in the high gain regime is not.

# 4   X-ray FELs

At the present time, the most intense sources of X-rays are undulators based on synchrotron radiation sources. Let us take as an example a storage ring like the APS at Argonne, with a beam energy of 8 GeV, average beam current of about 400 mA, and peak current of about 500A, with a pulse length of 20 ps, and a beam emittance of $2 \times 10^{-9} mrad$. Assume also that we use an undulator with $a_u = 1, \lambda_u = 1.6cm$ and $N_u = 100$ to produce radiation at $\lambda = 0.1nm$. Notice that since the emittance is larger than the wavelength the radiation production is diffraction limited. Using (14) we obtain an average and a peak brightness $B_{\lambda,ave} \approx 10^{20} ph/mm^2 mrad^2 0.1\%$, $B_{\lambda,p} \approx 10^{23} ph/mm^2 mrad^2 0.1\%$.

An X-ray FEL can provide many order of magnitudes larger peak brightness, because of the larger peak current and smaller emittance that we can obtain today using a new generation of electron sources, and the techniques for emittance

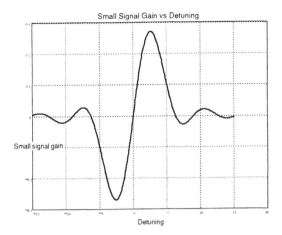

Figure 1:

preservation during acceleration developed for electron-positron linear colliders. As an example of an X-ray FEL we consider a possible system based on the SLAC linac; a list of beam and undulator parameters is given in Table 1.

| Energy, and relative energy spread | 15 GeV,0.0002 |
|---|---|
| Emittance, rms at 15 GeV | $3 \times 10^{-11}$ mrad |
| Peak current, and bunch length | 5kA, 24$\mu$m(80fs) |
| Undulator period, and undulator parameter | 2.67cm, 2.91 |
| Focusing wavelength in undulator | 38m |
| FEL parameter | $10^{-3}$ |
| Gain length, Saturation length | 4m, 42m |
| Peak brightness and Average brightness | $10^{32}$, and $10^{23}$ph/mm$^2$mrad$^2$/0.1% |
| Peak power | > 10GW |

Table 1: X-ray FEL list of parameters.

The possibility of operating an FEL at such short wavelength follows from the favourable scaling laws for this system. To obtain and discuss the FEL scaling laws let us write the FEL parameter using two quantities, the emittance, $\varepsilon$, and the longitudinal brightness, $B_L$, to characterize the electron beam. The longitudinal brightness is defined as (??)

$$B_L = \frac{eN_e}{2\pi\gamma\sigma_L\sigma_\gamma}, \tag{92}$$

where $\sigma_\gamma$ is the rms relative energy spread and $\sigma_L$ is the rms bunch length. The quantity in the denominator of (92) is the beam longitudinal phase-space area. Using the emittance and the longitudinal brightness the FEL parameter can be written as

$$\rho = (\frac{\sqrt{2\pi}}{8\pi^2}\frac{a_u^2}{1 + a_u^2}\lambda^2\gamma^2\sigma_\gamma\frac{B_L/I_A}{\varepsilon\beta_u})^{1/2}, \tag{93}$$

where $I_A$ is the Alfven current ($\approx 17kA$).

We also require that we satisfy the conditions for the validity of the 1-D model, equations (17, 18, 20), and assume $\sigma_\gamma = \rho/k_1, \varepsilon = k_2\lambda/4\pi$. The condition, (20), on the optical focusing can be shown to follow from the other two. We also use additional focusing to produce through the undulator a betatron oscillation wavelength of the order of the gain length, $\lambda_\beta \approx 2\pi L_G$. Using these conditions and rewriting the FEL parameter in terms of emittance and longitudinal brightness, we obtain (??)

$$\rho = \frac{1}{k_1 k_2}\frac{a_u^2}{1 + a_u^2}\frac{B_L}{I_A}. \tag{94}$$

Using typical values, $k_1 \approx 4$, and $k_2 \approx 6$ , we can see that to obtain a value of $\rho$ of the order of $10^{-3}$, the minimum compatible with a practical undulator, we need a longitudinal brightness of the order of 500A, a value which has been exceeded in photinjector electron sources. The scaling law (94) does not depend directly on the radiation wavelength, but only indirectltly through the requirements (17). This weak dependence of the FEL scaling law on the radiation wavelength is an important property, which can be used to develop an X-ray FEL.

## 5    Conclusions

We have shown that the main properties of an FEL are determined by the collective instability, and can be characterized by the FEL parameter $\rho$. The properties of the FEL, combined with recent progress in the production of high brightness electron beams, are such that this system promises to become a very powerful source of coherent X-rays.

## Acknowledgments

I wish to thank R. Bonifacio for all the discussions on FELs that we had over many years.

## References

[1] R. Bonifacio, C. Pellegrini, and L. Narducci, Opt. Commun., **50**, 373 (1984).

[2] C. Pellegrini, Jour. Opt. Soc. of Amer. **B2**, 259 (1985)

[3] K.J. Kim et al., Nucl. Instr. and Meth. **A239**, 54 (1985)

[4] K.J. Kim, Phys. Rev. Letters **57**, 1871 (1986)

[5] C. Pellegrini, Nucl. Instr. and Meth. **A272**, 364 (1988)

[6] R. L. Sheffield, Photocathode RF Guns, in Physics of Particle Accelerators, AIP vol.184, p. 1500. M. Month and M. Dienes eds., (1989).

[7] K. Bane, "Wakefield Effects in a Linear Collider", AIP Conf. Proc., vol. 153, p. 971 (1987)

[8] J. Seeman et al, "Summary of Emittance Control in the SLC Linac", US Particle Accelerator Conference, IEEE Conf. Proc. 91CH3038-7, p. 2064 (1991).

[9] M. Ross et al, "Wire Scanners for Beam Size and Emittance Measurements at the SLC", US Particle Accelerator Conference, IEEE Conf. Proc. 91CH3038-7, p. 1201 (1991).

[10] J. Seeman et al, "Multibunch Energy and Spectrum Control in the SLC High Energy Linac", US Particle Accelerator Conference, IEEE Conf. Proc. 91CH3038-7, p. 3210 (1991).

[11] T. Raubenheimer, "The Generation and Acceleration of Low Emittance Flat Beams for Future Linear Colliders", SLAC-Report 387 (1991).

[12] Workshop on Fourth Generation Light Sources, SSRL Report 92/02, 639 pages. M. Cornacchia and H. Winick, editors.

[13] C. Pellegrini, ibid, p.364.

[14] K.-J. Kim, ibid. p. 315.

[15] W. Barletta, A. Sessler and L. Yu, ibid, p.376-84

[16] C. Pellegrini et al., Nucl. Instr. and Meth. **A341**, 326 (1994).

[17] G. Travish et al., Nucl. Instr. and Meth. **A358**, 60 (1995).

[18] J. B. Murphy and C. Pellegrini, Introduction to the Physics of FELs, in Laser Handbook, Vol. 6, W.B. Colson, C. Pellegrini, and A. Renieri eds., North Holland, (1990).

[19] R. Bonifacio et al., La Rivista del Nuovo Cimento, **Vol. 13**, N. 9, (1990).

[20] J. M. J. Madey, J. Appl. Phys. **42**, 1906 (1971).

# INDEX

T - #0199 - 101024 - C0 - 229/152/9 [11] - CB - 9789056995027 - Gloss Lamination